# QUETZAL QUEST

Victor W. v. Hagen and Quail Hawkins

# QUETZAL QUEST

The story of the capture of the quetzal,
the sacred bird of the Aztecs
and the Maya

ILLUSTRATED BY ANTONIO SOTOMAYOR

London
OXFORD UNIVERSITY PRESS
1968

*Oxford University Press, Ely House, London W. 1*

GLASGOW NEW YORK TORONTO MELBOURNE WELLINGTON
CAPE TOWN SALISBURY IBADAN NAIROBI LUSAKA ADDIS ABABA
BOMBAY CALCUTTA MADRAS KARACHI LAHORE DACCA
KUALA LUMPUR HONG KONG TOKYO

First published 1940
Second impression 1948
First published in this edition 1968

For

ALEXA RIEGELS

who may some day see the bird
with the great green tail

*Printed in Great Britain by Richard Clay (The Chaucer Press) Ltd.,
Bungay, Suffolk*

# CONTENTS


1   THE ARRIVAL OF THE STRANGERS            1

2   THE SISIMIKI AGAIN                      15

3   TO THE NEST OF THE SACRED BIRD          21

4   FIDELIO FINDS THE SECRET                31

5   LIFE BEGINS AT EGG-HOOD                 37

6   THE QUETZALITOS LEARN TO WALK           49

7   THE ARMY ANT INVADES THE CAMP           62

8   THE SISIMIKI LEAVES ITS MARK            70

9   THE WHITE MAN DISPELS THE LEGEND        77

10  THE TREK BEGINS                         84

11  THE TREK ENDS                           92

12  THE QUETZAL MYTH IS BROKEN              96

    LIST OF SPANISH AND INDIAN WORDS        104

# FOREWORD

Dr. Hagen was the first man to bring the gorgeous and almost mythical Quetzal birds alive to Europe. You can see the birds he brought over in the Bird House at the London Zoo.* In *Quetzal Quest* he gives a vivid picture of the background of their life in nature, and indeed of tropical American life in general—the life of the half-civilized Indians, of the wild creatures in the mountain forests, of the traditions of the country's past greatness. He is to be congratulated on his skill as a writer as well as on his skill in finding and caring for rare wild creatures.

Julian Huxley

* *Note to Second Edition* According to London Zoo, the last of their quetzals died in 1967.

# THE ARRIVAL OF THE STRANGERS

It was July and nearly time for the corn harvest. The Indians of Portillo Grande went up every day into the hills to prepare their fields of corn, which they called *milpas*. The days were full of sun, and great white, billowy clouds marched across the heavens, casting moving shadows on the little Honduran village. Across the deep azure sky, the thick masses of clouds would settle into the green forests rising above the pine regions. Below this thick jungle the Indians had their cornfields. Almost any day they could be seen setting off for the fields at daybreak, all dressed in their short white cotton pants and shirt, with great high-peaked straw hats on their heads, and carrying their heavy shovel-like *pujantes*.

Old Chico, the elder of the village, was unable to join the younger men. He would sit by his small palm-thatched hut and smoke, pulling at the six or eight hairs on his chin, as he watched the other Indians scale the sides of the mountains, and saw them walking in their cornfields.

Fidelio joined the younger boys that day, for the corn-stalks had not yet been broken in half until the ears touched the ground, to keep the birds from eating the kernels. The boys, headed by Fidelio, had to sit on a small platform that the older men erected in the field. Shouting and making noises throughout the day, they scared away the birds that came to peck at the ripe corn.

Fidelio was a slight figure, small for his age, but active. During his twelve years he had not grown quickly, but Old Chico said that he was smart enough for one twice as big.

Sitting on the platform, above the rustling leaves of

corn, now and then giving a shout at the birds perched on the branches of the trees near by, Fidelio looked longingly at the thick jungle up and beyond the corn-fields. His inquisitive brown eyes were sparkling with interest.

From where he sat, he could see the white cliffs they called La Peña—the rock of sorrows, the mysterious place everyone spoke of with fear. Up there in those deep jungles were giant trees festooned with great, thick creepers. The jaguar lurked there and the

harpy eagle. Monkeys could be heard in the night of the full moon; howling monkeys that sang and shrieked throughout the night. There, too, was that beautiful bird with the green feathers that Old Chico said were like those once used for the crown of Quetzal-coatl, the Plumed Serpent. Fidelio wondered what he looked like. Old Chico said he was sometimes called the Fair God because he had a fair white skin and a black beard—so different from the Indians who worshipped him in long-ago Mexico. But he had gone away, on a raft of serpent skins, Old Chico had said, even before Moctezuma ruled.

Fidelio thought again of the night before by the fire. He could still see Old Chico putting his hand in the bin behind him and filling his straw hat with kernels of corn. These he had dropped, kernel by kernel, into the bin, saying, 'Let each one of these be a season.' Then he had let the whole of the corn in his hat flow like a golden stream back into the bin. 'Each kernel a season, and so many of these—that was the time when we were once great.' The Indians, who could not count beyond ten, had been astounded. They had never known anything as old as that. Old Chico, who was eighty years old, was the most ancient thing they knew. Fidelio shook his head in wonderment, even now.

Though the bright sunlight poured down on him, he shivered as he thought of what had happened when the story was finished. The big fire of pine logs had burned low, and they had been discussing the story. In his excitement, he had told them of the time he had gone up to La Peña, where, on the ground, he had seen the long green tail-feathers. He had not admitted watching the birds as well. He felt certain that what Old Chico had said was true, and that somewhere in the long distant past his people had gathered tribute for the great Moctezuma, the Emperor of Mexico.

'Do you suppose,' he asked himself, 'that my family once collected those green feathers for the emperor himself to wear? Just suppose!' He thrilled at the thought of those far-away days, when the Indians were important and had big fiestas and fairs.

A booming voice behind him jarred him from his reveries. 'Hey, Fidelio, you young fool! Do you see the birds eating the corn? Wave your shirt at them.'

Fidelio sprang up at Ignacio's harsh voice and shouted at the birds until they flew away. Still he could not keep his thoughts off La Peña. The bird with the green feathers—but the Sisimiki—oh, he shouldn't even think of that! The Indians said that if you thought of that beast, or even mentioned its name, it would come down from La Peña. Fidelio muttered to himself, 'I wonder what the Sisimiki is? Why is it so dangerous? Why did they scold me when I told them I had been up to La Peña?' He could still feel the cold, hard looks they gave him, and the way they avoided looking at each other, but stared uneasily into the fire. Old Chico had been angry and scolded him harshly, aghast at Fidelio's foolishness. Hadn't he told Fidelio never to go there? Did he want them all to die? The rest had all joined in upbraiding him for going up to that horrible place. They made him promise never to go there again, or even to mention the name of the Sisimiki. No one he knew had ever seen it, yet all said it was terrible. He wondered if it guarded the bird with the long green feathers.

As Fidelio sat thinking and thinking, the sun began to set. Already the other Indians were going home. He could see them swing the heavy *pujantes* across their shoulders, put the gourds over their arms, and walk down the steep path to their homes. Remembering that he, too, must hurry home, he scrambled down from the platform and ran quickly through the corn, falling over some of the stalks and losing his hat. He picked it up, panting in his anxiety to get to the open path. Not he, not Fidelio, would stay behind in the dark. Suppose the Sisimiki—no, he must not think of it. He must hurry! He sent himself flying down the slope.

Near the little village of Portillo Grande, which means the place-of-the-great-little-door, there is a path which leads, so Chico said, to a large village far down in the valley. That little road, wide enough for beasts, but for nothing else, clings to the side of the valley, and from the top of the slope you can see anyone coming for some distance.

As Fidelio neared the road he saw eight mules, some with loads attached to their backs. The others carried people. Who were they? They could not be Indians, for Indians don't ride horseback. Were they soldiers of the government coming to make them work on the road? Had someone done something wrong?

Fidelio hurried down and hid behind a large rock, where he could see the road, but could not be seen.

At last the figures came into view. No, they were not soldiers. Fidelio had never seen people like this before. When they came nearer, he saw that the man riding at the head of the party was large, his skin was white, and—no, not—it could not be true—he had a great black beard! The man was talking—what was he saying? Fidelio strained his ears, but he could make nothing of what they were saying. Fidelio could not believe his eyes—white skin—big black beard—spoke a strange language! Quetzalcoatl—the Plumed Serpent! Was the Fair God coming back? Fidelio could not contain himself.

He rose quietly and when he could not be seen, he ran as hard as he could to the village.

Chico was just putting the pigs into their pen for the night when Fidelio came running down the hill shouting:

'Quetzalcoatl is coming, Quetzalcoatl is coming!'

At the hysterical screaming of Fidelio, all the people of the village poured from their houses. The men and women gathered around Chico and Fidelio.

'What is he saying?' questioned Santiago. 'Who is coming? Quetzalcoatl, the Plumed Serpent? What is he talking about?'

'It's true, it's true!' panted Fidelio, still out of breath from running. 'I just saw them. Three people, one with white skin, a large black beard, speaking a strange language! It is Quetzalcoatl!'

Old Chico stood by, shaking his head slowly from side to side. He said nothing.

Fidelio tugged at his dirty, worn, cotton jacket. 'Isn't it true, Chico? Isn't it the Plumed Serpent coming back, as you said?'

A voice snarled from behind the group. 'So that's it—Old Chico has been telling the boy stories and he believes them! Maybe he has been bewitched. You know,' and Ignacio turned to those gathered about him, 'you know Fidelio has been to the mountain where the Sisimiki is——'

At the mention of that name, the Indian women put their hands to their ears and screamed and ran for their huts, their long, black pigtails streaming behind them like tails of a kite.

'Silence!' said Old Chico. 'We can't go now, for it will soon be dark. Tomorrow, when we have light, let us go and see what Fidelio has seen.'

The other Indians thought Fidelio was bewitched, and went complainingly to their houses. Old Chico, softened by the little boy's look of appeal, said, 'Come, Fidelio,' and, placing his arm about him, drew him into his hut.

At the break of day the next morning Fidelio, Old Chico, and three other Indians crept up the hill to discover if Fidelio really had seen something, or was suffering from some curse or enchantment.

Santiago went ahead, for he was more agile than the rest. He peeped over the top of the hill, and rose higher until his whole body was exposed. He stood for a moment watching, and then he made a motion for those below to come up. The others scrambled up the hill. Fidelio, however, waited to help Old Chico, who could not climb as fast as the others. Soon they joined the rest at the top.

There was a buzz of conversation. 'Who are they?' 'What are they doing?' 'So Fidelio isn't bewitched! He did see something!' Chico covered his forehead with his palm to shade his eyes from the rising sun. He stood there, motionless, his gaunt, bronze face showing the high, curved cheekbones, the black, straight hair that was brushed back from his forehead. Fidelio, looking intently at him, thought him very noble-looking. At last Chico spoke:

'Fidelio was right. One of them has a long black beard, and he is white; but I don't think it is the Plumed Serpent. They look more like gringos.'

'Gringos, gringos,' asked Fidelio, 'what are they? Are they Indians too, different Indians from us?'

'No, they are not Indians, *Jicaque*, like us.' There was a ghost of a smile about his lips. 'These are the white men I spoke of, that come in big rafts and have such curious things. I wonder'—and Chico broke off his narrative—'I wonder what they want here.'

In the clearing on the flat of the hill, Fidelio could see these strange people busily engaged, he supposed, in making a house for themselves. They dragged out a big bundle of cloth. One of them crawled under it, and then the thing raised itself. Some of the others got hold of the ropes, and soon the thing looked like a queer house. Yes, there was a door! The woman and one of the men went in. It looked like a house. But how quickly they put it up! Why, when Fidelio helped build their house of mud bricks and thatch, it took weeks. These certainly were strange people.

That evening, after Fidelio and the other Indians had finished their usual supper of tortillas and beans, the gringo and the gringo woman strode into their village. They were big people all right, thought Fidelio. He hid behind his house until they went up to the tumbledown hut of Chico. Then he heard them saying pleasantly enough:

'*Buenas tardes*—good evening!' Why, they spoke Spanish just like the *padres* and soldiers who came up! As they stood talking to Old Chico, all the curious people of the village came down—men, women, boys, and little girls. Fidelio moved about the house, and came over to Chico's side. His eyes, large and round, were filled with wonderment. The gringo had big arms, and—yes—his beard was black. He had a funny way of laughing deeply. But the woman was different. Her hair was the colour of corn silk next to the kernel, almost white. She had blue eyes instead of black. He had never seen anyone with eyes like that. The Indians, the soldiers, even the *padres*, had brown eyes. Perhaps she was blind. He once heard of an old Indian who had blue eyes, and he could not see. But no, she was not blind. She could see!

Old Chico, remembering his manners, the things that the *padre* told him, said: 'Fidelio, run inside and get the bench for these people.'

Fidelio, eager to be important, ran for the bench.

The man with the beard was talking now. At first Fidelio could not understand him clearly, although he knew Spanish almost as well as his Indian language.

The way the man spoke was short, crisp, and he seemed to cut his words. After a while Fidelio's ear became accustomed to the strange accent and he could understand.

'We have come here,' the white man was saying, 'to search for some rare creatures that we are told live in these mountains.' He waved his hand to the *montañas*, the thick deep jungles behind the Indians' *milpas*. 'We shall need some assistance; men for guides, and a boy to help in our camp, as the ones who came with us go back tomorrow. We will pay them well for their services.'

The gringos seemed at their ease, leaning back against the uprights of Chico's house. The white woman smoked as well as the bearded gringo. Presently he passed cigarettes round to all the Indians. As he came up to Fidelio, he smiled and said:

'And you, do you smoke, too?'

Fidelio was so frightened he could only say, 'No, señor,' and stare darkly at the ground.

Night settled down; the moon has not risen from behind the hills. The last rays of the sun caught the tips of the roofs of the houses. The fire in front of Chico's house was lighted, and the dry, pitch-filled pine spluttered and cracked as the flames leaped high. With the night came the mosquitoes. Big, black fellows that buzzed about the ears like bumblebees.

The white hunter slapped his face and then his ear, and laughingly said: 'Well, you have plenty of mosquitoes about.'

At this, Old Chico opened his mouth in a funny

cackle and laughed loud. '*Si, si*, yes, yes, that is one thing we have a great deal of here.'

Chico laughed long and loud at his own joke. He threw back his head and laughed with open mouth. Fidelio could see the one old tooth on top and the two others in the lower part of his mouth. Fidelio had always wanted to know how he could chew his tortillas with just those three teeth, when they didn't even come together.

It was obvious that the stranger had something else on his mind. Although he spoke of the long journey to the country, the cold nights here high above the sea, and the fact that this country, their country of Honduras, was not well known, Fidelio sensed, as did all the rest, that they would soon find out the real reason why the gringos had come. Perhaps for gold.

'We have come here,' said the hunter, as if in answer to the unspoken question, 'to find a bird which is called the quetzal. You perhaps know it. It is about the size of a large pigeon. The male has a crest, with a red breast and a very long green tail. We want to catch it alive. We would like you to help us. When we heard that the bird lived in these forests, we thought you must know it. Do you know the quetzal?'

Old Chico exchanged glances with all the other Indians about him. 'No, señor,' said Chico, shaking his head, his eyes now cold and hard. 'We have never heard of such a bird.' Like a chorus, the others gathered about said the same thing.

'But surely,' insisted the white hunter, 'you must have gone up to those mountains, up and beyond your cornfields, that I see in the distance. There surely you might have seen the bird.'

'No,' said another from the crowd of Indians, 'we don't go up to those hills. It is too far and there is nothing there like the bird you say, with a long green tail.'

Fidelio could hardly contain himself. He *had* seen the bird with the green tail, and he knew the others had also, although he did not know the name quetzal. Though that might be some other bird, there could be but one with the long green tail.

Now Chico was talking again: 'No, señor, we never go up there. It is true what Santiago told you. It is far and there is nothing there but trees and rocks. Not that we are afraid, but it's dangerous to walk up there. You might fall in between the rocks. So we don't know about anything.'

Suddenly Fidelio said, 'But, Chico, I thought it was because of the Sisi——!'

'Quiet, quiet!' shouted Chico, suddenly angry. 'You shouldn't speak.'

There was a moment of silence.

The two white people looked at each other and said something in another language and smiled, then to Chico:

'Well, it is of no importance. However, we would like to have one of the boys to help us in our camp, to get wood, to make the fire, and help about the house. We will pay him,' and the man reached in his pocket and took out five large pieces of silver, 'we will pay him this much each month and he can eat with us. If he wants to he can come home here to sleep.'

At the sight of the silver the Indians became interested, for whenever the officials of the government came, the Indians were never paid for their work. These people *were* different.

'Our names,' continued the white man, 'well—they are long. You may call my wife Doña Cristina, and myself Don Victor. Now let us see which boy will be best,' and he instantly eyed the intelligent face of Fidelio. 'How about you? Would you like to come and work for us? Come, come, you must not be afraid. We are just like you. We eat the same food. You'll come? Good! Come tomorrow early in the morning when the sun is so high,' and the white man stretched out his arm very low, the position in which the sun would be at about daybreak. For this is the way the Indians tell time.

2

## THE SISIMIKI AGAIN

How strange was this new world to Fidelio! True, he lived in the same place as he had always lived. From the camp of the white man he could just see the roof on his house, but here everything was peculiar. These people had all sorts of curious things: boxes and boxes of funny bottles, and things they called 'in-struments'.

Besides, these people were gay. They were not sullen like the Indians, who only laughed when you fell down and who never sang their mournful songs except when they were drunk. These white people were always laughing. They were good to Fidelio, too. They gave him things to eat that he never had had before.

He walked with the man in the forest and watched him pick up insects with little metal tweezers and put them in bottles. The white woman, too, did odd things.

She would go out in the forest with something she called a 'plant press' and into this, between thick paper, she would put the leaves of trees and their flowers, and ask all sorts of questions which she would write down. 'How big is the tree? What do you call it in Spanish, and what is it called in your language?'

All this puzzled Fidelio, but he grew to like his employers. After supper when they sat around in soft chairs, he would ask questions about the country where the white man came from. The white man, whom he now called Don Victor, would tell him of big ships. It was true what he had heard. They did have fire in the middle. But he heard more. About houses made of stone that were piled high one on top of the other, where as many as ten villages of people lived together. How surprised he was when he was told that people there do not always make fires with wood. 'Well, what do they cook with? Gas? What is gas?' When Don Victor tried to tell him, Fidelio said he understood, but really he did not. How could he? He thought Don Victor was telling him a funny story.

Then one day, Don Victor began to talk about the bird with the long green tail. He spoke of how he wished to get the bird, and that he had been up in the forest the other day, but saw none of them. Was Fidelio sure that he had never heard of a quetzal? Fidelio shook his head, for it was true; he had never heard of such a bird. But he knew secretly that it must be the bird that he had seen.

'Let me show you something,' and Don Victor got up to go into his tent. Meanwhile, Doña Cristina had taken something out of the oven—a big white piece of 'bread' which she had covered with something white on top.

She spoke to him. 'Have you ever eaten a piece of coconut cake?'

'No, señora,' he answered, 'I don't know what it is. I only eat tortillas and beans here, and sometimes a few peppers and maybe meat once in a while.'

Doña Cristina then gave him a large piece. It was all sticky but it smelled very good. He bit into it and decided he had tasted nothing like it before. It was soft, crunchy, and very sweet. He licked his fingers solemnly as he finished, his face beatific with enjoyment.

Don Victor came out of his tent bearing something in his hand. It was a wide, short, shiny knife in a new leather sheath which he laid on Fidelio's lap. The little Indian quickly wiped his hands, sticky with the remnants of the icing of the cake, on his cotton pants, and picked up the knife. He said nothing, but his eyes gleamed.

'How would you like that knife?' asked Don Victor.

'I would like it very much,' Fidelio managed to stammer out.

'Now this is what I want to know,' and Don Victor fumbled in a large book. 'Have you ever seen this bird before?' He placed a picture on Fidelio's lap. It was in colour: a bird with a great green crest, a red breast, two yellow feet protruding from the green plumage—and hanging from behind were six long green feathers.

Fidelio exclaimed: '*Ampusay!*'

'So,' Don Victor shouted enthusiastically, 'you do

know it! *Ampusay!* That's why you didn't know it as quetzal. Tell me, didn't Chico know what bird I was talking about the other night? You knew! For I saw that look come over your face.'

Fidelio was very embarrassed. 'I shouldn't tell you. Chico will be angry. Yes, he knew—we all know about that bird. And I think, although Chico never said so, that its feathers used to be worn by the Plumed Serpent.'

Don Victor was astounded. 'Where did you hear about the Plumed Serpent?'

'Chico told me,' said Fidelio. 'He tells me all about the Old Indians.'

'Well, then look here, I will show you that you are right,' and Don Victor brought out another coloured picture. It showed a tall, white Indian dressed with rich tunic, high sandals of jaguar fur, and a crown of green feathers.

'That's it, that's it!' shouted Fidelio. 'That is the Plumed Serpent. Just as Chico said!'

'Of course, of course! Now you see the connexion. The bird you called *ampusay* is the quetzal from which the Plumed Serpent gets his name, Quetzalcoatl—the serpent with plumes. Now what I want you to do is this: go up in the mountain and show me where you have seen this bird. Help me find it, and I shall give you many nice things. Besides, I shall pay you for your work. Perhaps you can get the others to work also.'

Fidelio became suddenly very grave. 'I can't do it. No! No! I can't! I am afraid of the Sisimiki!'

'The what?'

'Sisimiki, Sisimiki!'

'What in heaven's name is a Sisimiki?'

Fidelio was much upset. 'I don't know. I have never seen it, but I have seen its tracks in the forest. It lives up there in La Peña. They say it is about the size of a little child, walks on the ground upright, but looks like a monkey. It is Lord of the Jungle and when our Indians go up there to get thatch for our houses they always go two by two and never stay there after dark. They say it follows people, too.'

'Now, now,' said Don Victor, pursing his lips. 'That is interesting,' and he lapsed into his own language, ' "There are more things in heaven and earth than are dreamt of in your philosophy." '

'Now look here, Fidelio. We have guns, fine guns that can kill anything. We have lights that we can work with our hands. We are not afraid of the Sisimiki. White man has conquered more things than a little child that looks like a monkey and runs about in high, cold forests in the mountains. Did white man not come and conquer even the great Lord Moctezuma of Mexico? Well, now we would like to see, alive, this bird that you call *ampusay* and we call quetzal; and if we can get some alive, to bring them alive to our country. It never has been done. The bird dies in a few days when it is captured. It will not eat. So it is a great chance we are taking. If you will help me, I will do whatever you wish. You must see Chico alone. I want you to give him one of these knives. Tell him what I have said. Now he is your chief, your Elder, and if he asks the other Indians to help, they will. What do you say?'

Fidelio was aflame with the desire to help. He promised he would do what Don Victor asked and then said meekly:

'Did you say that you would do anything that I wanted, if I helped get the birds?'

Don Victor nodded.

'Well, maybe I could go down with you to the coast and see those ships you spoke of, and see what the white man does there.'

'Very well,' agreed Don Victor, 'it is done. You will see the ships and the white men on the coast.'

Fidelio went off to see Old Chico a very happy Indian boy.

# 3

# TO THE NEST OF THE SACRED BIRD

'But, Chico, Don Victor promised to give you this fine knife, if you will help find the bird he is looking for. See, it is sharp!' said Fidelio to his old friend and guardian, as he finished his plea for his new friend and employer.

'Yes, it is sharp—but what of the Sisimiki?' Old Chico was eyeing the knife possessively, but warily.

'Don Victor promised to carry a big gun to frighten the Sisimiki; and besides, he said if many of us went together, nothing could hurt us,' urged Fidelio.

'Who knows? Yet I think we go. The knife is good, and you say he promised much silver?' Chico turned over in his mind the desirability of assisting. Money was scarce, and always pleasant. Yes, perhaps all would be well. 'We will come at evening to see Don Victor.'

Fidelio hurried back to the camp triumphantly. Don Victor made plans to begin the hunt for the quetzal.

Shortly after sunset, a whole group of Indians, led by Fidelio and Old Chico, came into the white hunter's camp to see at first hand all the marvels that Fidelio had spoken of.

Yes, sure enough, there were the boxes, bottles, and shiny things he called 'in-struments'. Don Victor seemed very anxious to please Old Chico. He let him hold his binoculars to his eyes.

Old Chico completely forgot his dignity, and laughed and cackled until the tears rolled from his eyes. 'There he is, my old fat pig. So quick he comes when I look at him through the funny pipe.' He reached out to touch the pig still as far away as before, though it seemed under his hand. He looked surprised

not to touch the pig. Next, he fingered Don Victor's big shotgun, and rubbed his gnarled brown hand over the shiny, clean barrel.

'Is it not as I said?' demanded Fidelio excitedly, for he was proud of his knowledge of the white man and his ways. He made the Indians, who were shy and suspicious, as anxious to help as he had always been. They, too, wished to have shiny knives and silver pieces.

Still Old Chico was pessimistic. 'Don Victor, I have never heard of anyone keeping alive an *ampusay*.

Sure, sure, we have caught them alive. But they do not eat. It is no use! You will see.'

Don Victor looked very grave at this, and two big frowning lines came between his eyes. 'Well, Chico, why cannot we get the young birds? When do the *ampusay* lay their eggs?'

Chico raised his shoulder, stretched his hands at his sides in a gesture of complete lack of knowledge. '*Quién sabe?* Who knows?' None of the other Indians knew either, when or where the *ampusay* laid its eggs. Nevertheless plans were made for a great hunt to begin the next day.

Early, almost at sunrise, the hunt began! Don Victor told all the Indians to be sure not to kill the *ampusay*. If they should, and it was egg-laying time, they would lose some of the birds they might have had.

Ten of the Indians, including Beltran, Santiago, Pedro, and Pepe, went up to the forest, moving off towards the great bare rock of La Peña. Don Victor, with Doña Cristina, Fidelio, and another Indian called Miguel, went off in another direction into the *montaña*. It was the first time Fidelio had ever been in this section.

'It is a long climb, señor,' he said seriously to Don Victor as they panted up the slope, higher and higher, leaving the cornfields behind. Don Victor and Doña Cristina, who did not seem to be used to climbing, would stop and puff out their cheeks, perspiration rolling down their faces, but their anxiety to find the quetzal made them push on and on.

At noontime, after climbing four hours, they entered the cloud forest. They were so high that the ordinary white clouds which Fidelio always saw over-head, big white masses like the feathers of a hen, now lay right in front of him like fog.

'Why, I could go right up and take a handful of cloud. It is curious,' he said to Doña Cristina, as she watched him walk towards a bit of cloud drifting closer to his hands. When he opened them there was nothing there. 'Ai! where does it go, that cloud? First it is here and then it vanishes!' and Fidelio laughed, showing his even white teeth.

Don Victor and Doña Cristina sat down to smoke while Fidelio looked about him. Miguel squatted in-differently by. Nothing surprised him. He was as stolid as most Indians.

Fidelio was more lively. 'Such big trees! Why, if ten Indians held hands and formed a circle they could not reach around that tree! And so high!' Fidelio leaned back on the log he was sitting on to see to the top—and, thump, he fell over. Even Miguel laughed at this mishap. As Fidelio lay where he fell, he saw some-thing moving above. Oh, they were monkeys, and he pointed and shouted, 'Look, Don Victor, monkeys! Do you see the big black one with the baby holding on to its fur?'

Don Victor jumped up and, pointing a black box at them, held it a while. When he heard Doña Cristina say, 'Did you get a good shot?' Fidelio thought it might be a kind of gun, and held his fingers to his ears.

As she noticed his screwed up face, she laughingly took Fidelio's hands from his ears. 'Don't be afraid, that is not a gun. It is a camera.'

'A ca-ma-ra,' repeated Fidelio. 'What does a ca-ca-ca-ma-ra do?'

She tried to explain. 'It takes a picture and keeps for ever what we see for a moment only. Do you understand?'

'*Si*, señora,' nodded Fidelio seriously, but he hadn't the faintest notion what she meant. Some magic, no doubt.

They ate their lunch as they rested, then started on, walking through the cloud jungle. The mist moved through the trees like the ghosts Old Chico used to tell about. The trees were not like the pine and oak below. Swinging in a tangled mass from tree to tree were vines as round and thick as Fidelio's wrist.

'Why,' thought Fidelio, 'I could climb on those just like the monkeys.' He decided not to try it just then, but later, perhaps, who knows?

The trees were covered, too, with other plants that were stuck to the branches and boughs. Some of the great moss hung down like long grey beards.

'It is silent here, señor, no?' murmured Fidelio a

B

bit nervously. The only noise that could be heard was the cracking of the twigs under the boots of the white hunter and his wife. Fidelio made no noise, nor did Miguel. They walked ahead, gazing in the air, looking for the bird with the long green tail. Miguel carried a blowgun. It was a pole about six feet long, and hollowed out inside. In a skin bag around his waist he had whole handfuls of clay-like pellets. 'They look,' said Don Victor as he noticed them, 'like marbles which the boys in my country play with.'

With this gun, Miguel, who was much older than Fidelio, could shoot birds or monkeys with just as much skill as Don Victor could with his shiny one. Fidelio liked Miguel, for he would show him how to do all those things he had always wanted to know.

Small deer, who were as curious about these intruders as the Indians were about the white hunter, would stand in the path, raising their quivering noses and trying to get a scent of the moving things. When Fidelio came near, off they would go in great bounds into the jungle.

Miguel was some distance ahead in the jungle path. He stopped short, and raised his hand as a signal for everyone to be quiet. Don Victor and Doña Cristina looked very curious. Fidelio stood as if frozen, his brown eyes gleaming with anticipation.

The silence was broken by a bird's voice, 'Whe-oooooo, whe-oooooo.' It was a soft double note. Then the 'whe-oooooo' took on a stronger sound, gathering volume until it was very loud. There was again silence in the jungle. Soon, far off, Fidelio could hear the notes again—'whe-ooo,' very softly, and as it grew louder and louder, other voices joined in.

Don Victor could keep still no longer, and tiptoed up beside Miguel, whose head was twisted to one side,

like a dog who wants to hear better. 'What is it?' he whispered.

'*Ampusay! Ampusay!*' Miguel whispered in reply.

'Quetzal,' translated Fidelio.

'Where?' asked Don Victor as Doña Cristina crowded close by.

'Up there—somewhere—in that *aguacatillo* tree, the tree of the little avocados,' said Fidelio, pointing. 'See the birds eating the fruit? No! No! This way! Look, there is one!'

Don Victor could see nothing, his eyes not being trained like the Indians' to see things in trees.

'Watch, now,' said Miguel, and raised his blowgun to his lips.

'Don't kill it,' pleaded Doña Cristina.

Miguel shook his head, saying, 'Watch.'

He took a deep breath until his lungs seemed to be bursting, and then blew. Fidelio could not see the pellet, but he heard it crack against the limb of the tree in the high boughs.

At that noise, the birds, which Don Victor had not seen, took flight. They were quetzals, six of them, all

male birds with long green tail feathers trailing behind them like strands of spun gold being pulled through the air.

Don Victor just stood there with his mouth open. Tears gathered in Doña Cristina's eyes, she was so happy.

'Just think,' Don Victor said to his wife, 'we have actually seen a quetzal bird. They are really here! Aren't they the most beautiful creatures you have ever seen?' She nodded her head, too overcome to speak.

Fidelio, who had crept forward, came back. 'Don Victor, look! It is sitting low on a tree.' He led them to a tree where, not twenty feet from the ground, a great male quetzal, with flaming red breast, was climbing to a bough, and he was singing. As he whistled, 'whe-oooo, whe-oooo', the long tail would flick up and down just like a happy little puppy's.

Again Don Victor took out his camera, crept very close, and took the bird's picture. Fidelio did not know which was the more curious, the white hunter or the quetzal. Each stared at the other. At last the bird took flight, and disappeared into the foliage. It faded right into the leaves, for the green on its feathers was the exact colour of the trees.

Don Victor could talk of nothing else in his excitement. His face wore an almost constant smile, and Fidelio knew why. He had come from far away, too far even to tell Fidelio, hoping to see these birds with the long green tails, and now he had seen them.

Fidelio was pleased that he had brought it about. Even though they saw no more that day, Don Victor was gay. That evening at camp he whistled, and made big plans for the morrow. What if the other Indians said they had seen none, he, Don Victor, knew they were in the great rain forest. Nothing would stop him. Fidelio watched him happily. He had helped Don Victor. Who knew, perhaps he might find a nest!

4

# FIDELIO FINDS THE SECRET

Coming to a fallen tree, Don Victor sat down. He said, 'You and Miguel go on. We are going to change film.'

'Change fil—change film?' queried Fidelio.

'Yes,' Don Victor said. 'I have taken all the pictures I can now. I must put more film into the camera.'

Fidelio shook his head. These white men were funny, and he muttered to himself as he went after Miguel, 'Change, fil, change fil—mm,' and shook his head sadly. There is much for an Indian boy to learn.

They were out again in search of the elusive quetzal. For many days now they had made daily expeditions, and had seen several birds, but had found no nests. It was important to locate them, for Don Victor felt sure it would be nesting time. He wanted to see the baby birds, the quetzalitos. No one knew where to look for nests. There were stories of what they were like—with a hole at both ends to take care of the tail— but no one had seen any. Fidelio wandered on, looking hopefully in the air for some sign of a nest.

Without realizing it, he had lost Miguel in the forest. He waited a moment and shouted, 'Miguel, Miguel!' But there was no answer. He looked around to see the way he had gone, and searched for his own footprints in the black, moist soil of the jungle. It was drizzling now, and the afternoon mists were coming down.

Suddenly he stopped short. His eyes, glued to some marks in the ground, took on a long look of horror and unbelief! There, there, right on the ground, as fresh as if it had just passed by, were the marks of a foot, a tiny foot like a young child's! The mark of the Sisimiki! There was no doubt—only the Sisimiki made prints like a child!

Beads of perspiration stood out on Fidelio's forehead. He looked fearfully over his shoulder. The mists closed in, intangibly surrounding him with their moist vagueness. Terror-stricken and dreadfully afraid, Fidelio ran. He didn't know where, he didn't know how. Stumbling, he crashed through the jungle as if all the wild animals of the world were at his heels. He tripped over vines, got up, only to fall over the wide buttresses of the trees.

Gasping, out of breath, and his fear somewhat allayed by his flight, he finally came to an opening in the forest. In the past, the wind must have blown down a big tree that dragged down many others in its path, for they were all tied together with those vines on the floor of the jungle. In the midst of this open space was a single dead tree. Fidelio leaned against another facing it, and closed his eyes for a moment, his chest rising and falling as he caught his breath, and his heart beating so hard it sounded like the 'in-stru-ment' Don Victor called a 'clock', which Fidelio had once held to his ears.

Suddenly the sun pierced the fog, and the open spot in the forest was filled with a shaft of light. As Fidelio

stood with his head thrown back, his eyes half open, he could see through the gratings of his long eyelashes. Something caught his eye, like the gleam, the reflection of the sun on his knife. He opened his eyes wider. In the dead tree, about thirty feet from the ground, was a hole, and from it four long green feathers hung. Swaying in the slight breeze, they caught the rays of the sun and sent out shimmering gold sparks of light. Why, it was the tail of a quetzal—and the hole—it must be the nest!

Fidelio moved forward. His foot caught on a dry branch, which cracked with a noise like Don Victor's shotgun when it went off. The tail disappeared, and in its place came a head—the head of a quetzal. It filled the whole entrance to the nest. The green crest stood up and Fidelio could just see the quetzal's breast, blazing red. The bird eyed Fidelio for a moment, then took an easy flight to a tree in the green jungle and watched silently.

Suddenly, Fidelio's mind began to function. Why! He had found the nest of the quetzal! He had forgotten the Sisimiki in his astonishment. He raised his voice and shouted, loud and long: 'Don Victor, Don Victor, come here, come here!'

Soon he heard an answering shout from the depth of the woods, and Don Victor came crashing along, making a noise like a big cow going through the high weeds near the river. The white hunter came, gun in hand, ready to use it.

'What is the matter, Fidelio? I thought you were being eaten alive!'

Fidelio rushed up, saying incoherently, 'I have found—I have found—look, what I have found, señor. See, it is the quetzal's nest!' He pointed to the dead tree with the hole.

Don Victor walked over to examine it. 'It looks more like a woodpecker's nest. How do you know it is the home of the quetzal?'

'Because of the feathers—the long green feathers— they hung out like the pigtail on Miguel—out from the nest. Then when he heard me, he pulled them in, and looked at me, like this, with his crest up, as if he were surprised to see Fidelio.' Cocking his head on one side he tried to show Don Victor how the bird looked at him. 'Then he flew out of the nest, and there he sits, señor,' and he pointed to the quetzal still sitting in the tree near the nest, waiting and eyeing the intruders, undisturbed by their excited chatter.

'And now,' said Don Victor, catching Fidelio's fever, 'if there are eggs inside the nest——!'

By this time Miguel and Doña Cristina had come into the bare patch and were told of Fidelio's find.

Eager to discover whether they would find eggs,

Don Victor went on. 'Quickly,' he ordered the Indians, 'take your big knives and go into the forest. Cut some slender trees and vines. Fix up a platform so that Fidelio can climb up to see inside the nest. Hurry!'

Soon they were back, dragging long slender poles behind them. In a short time a shaky structure, strong

enough for Fidelio to climb, was constructed. The quetzal bird, meanwhile, tired of waiting for them to leave, gave a small screech and flew off into the deep forest.

Fidelio climbed the platform like a monkey. He grasped the poles, lifting himself higher and higher, until at last he could look into the nest. The decayed old tree, with the added weight of the platform and Fidelio, swayed dangerously.

'Hurry, hurry,' said Miguel, 'or the tree will fall over.'

Now Fidelio was up to the nest. He put his hand in

and turned around, looking down, puzzled. 'It is very deep. I cannot feel anything.' He climbed farther up until he could put his whole arm in the hole. He reached in, deep down, up to his shoulder. Everyone below was breathless.

Then Fidelio pulled out his hand. In his dark brown fist were two eggs, smaller than chicken's eggs, but a pale blue, as azure as the sky.

'Fine, fine,' Don Victor said. 'Now put them back again.'

'But,' protested Fidelio, 'don't you want the eggs?'

'No! no! Put them in quickly, carefully, the way you found them.'

Fidelio did so, thinking Don Victor very strange. Here, he, Fidelio, had found the eggs Don Victor wanted so much and now he did not want them. He climbed down from the shaky platform a very puzzled, unhappy boy.

# 5

# LIFE BEGINS AT EGG-HOOD

As she noticed his puzzled expression, Doña Cristina ran over to Fidelio and caught him affectionately to her side. 'It was very clever of you to find that nest. Don Victor and I are proud of you! I shall have to make you another big coconut cake.'

'But why did Don Victor not want the eggs?' Even the promise of coconut cake could not turn Fidelio from his woe.

'Don't you know why?' asked Doña Cristina. 'He wants the mother bird to hatch out the eggs.'

'But then the quetzalitos will fly away,' protested Fidelio.

'No, no, they will not. Don Victor will not allow it, you will see. He will show you what to do,' Doña Cristina comforted him. He began to cheer up. Of a certainty Don Victor would not allow it.

On the return to camp, they found the other Indians waiting for them. Chico said they had seen many birds but no nests. After all, perhaps they had no nests. Perhaps they sprang full-fledged from the tree-tops. Who knew? The other Indians nodded in sullen agreement. It was fool's work to try to find a nest you could not see.

'Ho,' boasted Fidelio, full of his exploit. 'I know what their nests look like. I, Fidelio, have found a nest.

It stands high in a dead tree and has a little round hole, just big enough for the bird to enter. I saw the long green tail in the sun. It is easy to find. You will see!'

Don Victor said, holding two large pieces of silver in his hands, 'I will give these for each and every nest you find.'

Chico wanted more information. 'And then, Don Victor——'

'Then,' continued Don Victor, 'I want you to make sure there are eggs within. After you are sure, you will build a tall platform level with the nest, like the ones you make for the little boys in the cornfields, to scare away the birds.'

'You wish to scare away the birds, señor?' Santiago asked, somewhat at a loss. 'Why find the birds if you wish to frighten them?'

'No, no, Santiago, I do not wish to frighten the birds. Make the platform big enough for me to sit on top. I wish to watch the *ampusay*.'

Day after day the Indians, helped by Fidelio, went into the high rain forests searching for the nests of the *ampusay*. Each day brought new discoveries, until nine nests had been found. Under the direction of Don Victor the Indians raised observation platforms before each nest, while Fidelio, the most experienced of all the bird hunters, would accompany Don Victor to the forest to watch the birds.

'Why must we sit here and watch?' he asked one day, as they moved towards the platforms.

'Well, Fidelio,' and Don Victor placed his hand affectionately on the boy's shoulder, 'we must find out what they eat. We must watch the parent birds to see what they bring the nestlings, for no one knows what the quetzal feeds on. If we really want to raise them, we must give them the proper food. Now, you sit on top here,' and Don Victor pointed to a wooden frame tied together with the jungle vines. 'Climb up and sit quietly. I want you to see when the father bird comes to the nest, when he flies away, and when the mother bird comes.' Don Victor went off to another part of the jungle to sit also on top of a platform and observe his birds.

This was not much fun. It began to rain more often, for it was the beginning of August, and the *chubascos*, the great hurricanes, would soon come sweeping up from the east. Although Fidelio did not enjoy this part of his labour, he supposed Don Victor knew best. For long days he would sit watching the birds.

It was quiet in the great rain forest during the day. The sun hardly penetrated. Stuck to the sides of the trees, great plant parasites dripped moisture throughout the day. Now and then a large wasp whirled up in search of food. Shiny golden beetles would fly by with a *brrrrrr-rrr* and disappear in the foliage. Most of the animals came out at night. The daytime belonged to the insects. Spiders mended their nets. Ants skirted quickly about, tapping everything with their antennae. Other ants, long, black, marauding ants, went through the forest in black, solid phalanxes, sending other insects flying in dismay in front of the ant-hordes. With only these exceptions, life on the platform was as dull for Fidelio as it was sitting below, chasing the birds away from the cornfields.

When he arrived at his platform in the early morning, the green tail feathers of the male would always
be hanging out of the nest. As soon as Fidelio appeared, the quetzal would draw in the tail and look
the intruder over with a cold eye. When he was satis-

fied that Fidelio had no intention of leaving, he would
turn around, and re-enter the nest to sit on the two
blue eggs, his tail hanging out again.

During the day he would leave for short intervals in
search of food. At dusk, the female would return, very
quietly. She would sit on a branch near the nest, and
then, when she felt everything was safe, in one fast
swoop she would fly to the hole and enter. About this

time, Don Victor would come along carrying his long
gun, and Fidelio and he would return to camp together. They were always wet by the time they got
back, for it rained very hard at night.

Supper would be ready when they arrived, hungry
and wet. Doña Cristina and her Indian girl servant
always had a good hot meal of tortillas and beans and
sour cream, and onions and red peppers, finishing
with cake and coffee. Fidelio ate until he could hold
no more. Such good food these people had! After-
wards, Don Victor would tell tales of the wonderful

place he came from called 'U-nit-ed St-ates', until Fidelio was fast asleep.

Fidelio had just become used to the routine of watching the quetzal when something happened. He had mounted the platform as usual, and settled down quietly. The green tail of the quetzal hung out. After a while the bird left the nest and flew to the forest. Fidelio heard a curious noise, a *sssssssssss* like steam escaping from the pots that Doña Cristina had on her stove. Fidelio thought it was a snake, and grasped his big jungle knife tightly. Though he looked carefully about him, he could see nothing strange. Then the mother quetzal flew to the nest. This was odd because she never came back until evening, and it was hardly noon.

No sooner had she perched on the nest than the hissing became louder. She disappeared and the hissing sound died down to a faint splutter. The father bird came back, and once more the *sssssssssssss* grew loud. Fidelio was too startled to guess what could be happening. Perhaps a snake was in the nest. He must get Don Victor quickly. He slipped from the platform and ran lightly to Don Victor, who was sitting watching another nest.

'Don Victor,' he called, and made a motion to come down. The white hunter came down very silently. He was much more agile than he used to be. At first when he walked in the jungle, he knocked into every tree in his path, but now he was almost as quiet as the Indians.

'What is it, Fidelio?' he asked.

'I do not know. Something strange has happened.

The nest makes noises like a snake. Perhaps a snake has eaten the eggs.'

They hurried back to Fidelio's stand. Don Victor listened. 'I wonder if that could be the quetzalitos, the baby quetzals, coming out of their eggs. It should be time. I believe it is!' and a smile came over his face as he took from his pocket a small black book in which he wrote.

The next week was a busy one in the white man's camp. Don Victor made several small cages of wood with mosquito netting about them. As he had discovered that the mother quetzal feeds the little birds on the fruit of the *aguacatillo*, Fidelio had to scour the countryside to find enough to have plenty of fruit ready when they brought the quetzalitos to camp. It was exciting, too, for soon they would know if they could keep the birds alive. Chico and the other Indians were sure they would die.

There were still days when Fidelio had to sit upon the platform watching the mother quetzal bringing food to the little birds. The hissing increased. A tremendous fuss was going on inside the nest. Fidelio watched it for more than nine days after the birds hatched, but never once had he seen a baby bird. He wondered what they looked like. As if in answer to his question, the next day, while he was sitting in his accustomed place, Fidelio saw one little fledgling fly up to the hole of the nest, catch hold of the edge with its beak, and sit, blocking the entrance.

Fidelio was surprised. It did not look like a quetzal at all. It was covered with a grey and brown nest-down, and was fat and fluffy like a ball of brown cotton-wool.

Full of curiosity, it stood looking down on the world. Then it began to stretch its wings, balancing itself on the edge. First one wing, then the other! Suddenly it flew straight forward, trying to fly towards a branch. Striking the branch a little lower than it should, it struggled to lift itself up with its beak. At last, unsuccessful, it fell fluttering to the earth. But it was not hurt. Although it tried to fly again, flapping its wings helplessly, it could not rise off the ground.

Fidelio watched for a moment; then as he began to climb down to get the quetzalito, a dark shadow flew by. Something else had seen the little bird. A great grey and black bird came to rest on a limb only a few feet above the little quetzal. Fidelio saw it was a harpy eagle, the most feared of the birds of the forest. High up on the treetops it sits and watches. Anything that moves it glides down to attack. As Fidelio picked up a stick and threw it, the grey, rapacious bird of prey flew high in the air. Picking up the little bird, Fidelio climbed to the nest and put it gently inside.

When he told Don Victor about it, the white hunter made a quick decision. 'We must remove the birds tonight. Fidelio, go down and tell Santiago, Beltran, and Miguel to come tonight with pine torches, for we must remove the birds before they fly.'

The Indians were afraid to go up in the *montaña* at night even with the white hunter. At last he gave them some additional pieces of silver, but they were still afraid.

'All of us together will be safe,' pleaded Fidelio. 'I, myself, with Don Victor have been in the *montaña* at night, and nothing happened.'

Reluctantly and sullenly they agreed to go, and Fidelio knew they were all afraid of the Sisimiki. He was afraid, too, but Don Victor was not, and he would protect them.

They waited until quite late, until the mother quetzal would be in deep sleep before they went to the nest. As Don Victor had decided to remove the birds from three of the nests, these would all have to be visited. The Indians arrived with their torches alight, the burning pine spluttering and sending off bits of flaming rosin.

'Looks like fireworks,' said Don Victor, who was leading the way carrying an electric flashlight.

The party set off. All the Indians carried torches, except Fidelio, who was in the centre carrying Don Victor's big shotgun and some small baskets.

'We'll carry the gun,' said Don Victor, 'to keep the other Indians from being afraid of the Sisimiki.'

Fidelio was reassured. He felt none too safe himself, in spite of his boasting to the others.

Far ahead, the Indians began to climb, and as they moved along in the black night, it seemed to Fidelio as if they were the great lightning bugs that fly in the time of the corn harvest.

Oh, how dark the night was! The torches cast moving shadows in the jungle, and Fidelio out of the side of his eyes could, or at least thought he could, see all sorts of horrible animals in the dark. A great owl near them gave a loud, startled hoot—and Fidelio jumped, almost dropping Don Victor's gun.

They went very quietly, so as not to disturb the birds. Under the first nest, they held a conference.

'Now, Santiago, you climb up with Fidelio. You hold this flashlight right up to the nest, turned on until it is in the eyes of the mother quetzal and she cannot see. Fidelio will slip his hand about the big quetzal, hold her tightly and put her into his basket. Then lower it to me.'

Both Santiago and Fidelio nodded. Don Victor was not satisfied, however, until he made them repeat just what they were going to do. Then up the tree they went, the others bracing it from below. When they were half-way up, the quetzal appeared at the entrance to her nest, but as Santiago kept the light full in her face, she sat there, dazed, and made no attempt to fly.

Quickly Fidelio came close enough to seize her. He held her tightly to keep her wings from moving, and then, without letting go, he put her in the basket that Santiago handed him. Tying a cloth about it, he lowered it. Detaching one basket, those below put on another and Fidelio pulled it up again, to dip into the nest, this time taking out one by one the two fat little nestlings. The birds, very puzzled, opened their eyes wide and stared, but when Fidelio put them into the basket, they snuggled up close together and went to sleep again.

The second nest was visited without mishap. They now had two big birds and four nestlings. When they came to the third nest, Don Victor, who was not used to moving in the jungle at night, tripped and fell, making a loud noise.

Quickly flashing his light on the nest, Fidelio could see the female sitting on the edge, intense and excited. One move below, and she flew off into the forest. For some reason there was only one little bird in the nest, which they took.

Don Victor was happy. Fidelio could tell by the way he kept talking to the little birds, five little quetzals all huddled in the basket. The Indians, too, were huddled just as close, afraid to move out of the light of the torches. They kept looking fearfully over their shoulders into the shadows. Fidelio held the white hunter's gun, and stood watching Don Victor as he packed the birds for their trip down to camp.

Suddenly the acute eyes of the little Indian saw something moving towards Don Victor, sliding right into the arch of light of the flashlight. Fidelio tried to scream, but he could not. Lifting the big gun to his shoulder as he had seen Don Victor do, he pulled the trigger.

Booommmmmmmmmmmmmm——

The kick of the gun knocked Fidelio into a complete somersault through the air, and he landed on his face. Don Victor did not know what had happened. The shot had come so close to him that he had jumped back. He was about to upbraid Fidelio for being careless with his gun, when he saw the remains of a snake before him. It was the head of the terrible 'X' snake, the *Barba Amarillo*, its mangled body as thick as Don Victor's arm. What was left of it still wriggled, and its mouth opened and closed in the convulsions of death. The great fangs that dealt nocturnal death were stretched out.

Don Victor swallowed hard. 'Fidelio, if it had not been for your sharp eyes—Are you hurt?'

Fidelio rose, rubbing his shoulder where the gun had bruised him from its kick. 'I feel the way I did when Old Chico's pig ran into me.' They all laughed, the tension broken.

# 6

# THE QUETZALITOS LEARN TO WALK

Down in the village, Fidelio was telling Old Chico about the quetzalitos. 'They eat, those quetzalitos, all the time. Ten, even twice ten times a day, Doña Cristina goes to their baskets to feed them.'

'As many as that?' asked Santiago, who, with the others, was eager to hear about the babies. 'Do they never stop?'

'They hiss and scream for more. They are never satisfied, those quetzalitos. I think their bellies will burst, they are so big.' Fidelio rubbed his own fat little stomach as he spoke.

'They are not dead?' asked Miguel.

'What do they eat?' Chico asked, looking scornfully at Miguel. 'If they are eating, how could they be dead? But they will be dead. Soon Don Victor will see.'

'Avocados and eggs. Don Victor wishes you to find more for him. He will pay you,' answered Fidelio.

'How much does Don Victor pay you?' asked one curious Indian. The questions continued fast and furiously until Fidelio was glad to gather his bundles and hurry back to camp.

The white hunter was odd, he thought as he returned. After getting the big birds, he did not keep them. At first he kept them in small cages, but the

quetzals would not eat. Not one piece of avocado.
When Don Victor let them go, Fidelio did not mind.
He did not like them; they were females and had
no long tails. Oh, they were pretty, all right, but
not like the males, with the long tail of green feathers
which the Mexican chieftains used to wear. The female's tail was short and black and white and green.
Instead of having a blazing red breast, she has one of
grey, and she had no big crown like her husband.

Nothing would induce them to eat. Don Victor
made Fidelio hold each bird while he fed her by
stuffing the food down her throat. But even that was
not successful.

'I think we shall have to let them go, Fidelio. Now
you open the cage, and see if she will get on your
finger. I want to take her picture.'

Fidelio put his hand cautiously into the box, thinking the bird would peck at it, but she just sat there
looking very dignified and haughty. When he put his
finger against her legs, she got on it, and sat there. He

took her out into the open. Still the bird sat there, looking around. She was not going to be fooled. The quetzal is too dignified for that. If they wanted her to fly away, she would, but she would take no chances of trying and failing.

Meanwhile Don Victor was taking pictures with his 'ca-ca-ma-ra'. He came close; he sat down; he lay down; he stood up.

'This white man is strange,' thought Fidelio, watching. 'Every time he has the little black box in his hand, Don Victor does these queer things.'

At last the quetzal had had enough. She took off in swift flight for the forest and did not stop at all to look at her young sitting in a basket not far away.

Soon the little quetzals would not stay in their baskets. They were growing up, and did not belong in the baskets like nestlings. Don Victor cut the lower part of one wing, which prevented them from flying a great distance. They would have freedom of a sort.

Other birds were brought in, and before long there were twelve fat little quetzals—some big, some small, but all the same grey-brown colour.

'Don Victor, what shall I do?' Fidelio asked one morning as he kept thrusting the quetzalitos back into their baskets. They hopped to the edge as quickly as they were released.

'It is time to put them in their big cage,' Don Victor replied. The big cage, made of mosquito netting, allowed them room to jump about.

'Look, Doña Cristina, they are growing up. See the green coming on their backs! There are more green feathers here and here and here!' Fidelio pointed proudly at the rapidly changing birds.

Even so, Fidelio thought, there was too much fuss about them. Doña Cristina insisted on having him clear their cage three times a day, and three times a day he had to take them out of the cage for exercise. That he did not mind. But cleaning! Foolish! Did they not soil it again at once?

During their exercise period Fidelio taught the birds to get on his finger as he had trained his parrot to do. Then, one by one, he would put them outside the cage. They would fly a little, and then try to run on the ground, although they could not.

Don Victor explained why. 'You see, Fidelio, quetzals cannot run on the ground. You have seen that they never come down to the earth, but stay in the trees. They cannot run because their feet are formed only for grasping the round limb of a tree.'

'They are too proud to come down,' beamed Fidelio as if he had made a discovery.

'Perhaps,' agreed Don Victor, 'but we have to train them to run.'

'Why must they run?'

'Because if we don't the birds will not get their exercise. That is why all the other people could not keep the quetzal alive. They just put it into a cage and señor quetzal would be sad and lonely. He could not fly. He would not eat. He would die!'

'If we make them walk, they will not get sick and die?' Fidelio wanted to know.

'Perhaps.'

Fidelio, as he played and worked with them, grew

to love the birds. Each one had its own personality, just as Santiago and Beltran and Miguel and Old Chico had. Some were fat, and some were not; some hopped, others flew; some liked avocado, some egg, some *massa* of tortilla. Fidelio could tell them apart, and knew the characters of each. They were just like people. He explained it to Doña Cristina. 'Old Chico does not like to have you poke his pigs, and this fat little quetzal,' pointing to the one in his hand, 'won't let me fuss with his feathers to look for the green ones coming in.' Doña Cristina agreed, and looked more closely at the observing little boy. She was very fond of him, and found him quite different from the average Indian, who is inclined to notice little and express less.

The best fun Fidelio had was rolling small pieces of ground corn on the floor. He thought of this idea himself. At first when he put the birds on the floor they would just sit there. He tried to make them move by giving them a little push from behind, but as they were not able to walk, they would fall forward on their beaks. One day he took some ground corn from the kitchen and made it into balls. When he put the quetzalitos on the floor, he rolled the balls of corn along in front of them. Ah, that caught their eye, and with short, quick steps the quetzalitos hurried after the corn and snatched it up. Soon they were able to walk with greater ease. Don Victor thought that fine!

'They will be able to eat for themselves, before long, instead of being fed with a spoon,' he remarked one day, watching Fidelio with them. 'How old are they? Three weeks? Splendid! Fidelio, I think we shall succeed, and you shall have your trip to the coast.'

'To see the ships?'

'Yes, to see the ships. Now let us put the quetzalitos back in their cage and I will tell you something.'

Fidelio came back soon, the birds safe.

'In olden times,' Don Victor began, 'after the Plumed Serpent had gone, there was a great Mexican emperor called Moc——'

'Moctezuma,' exclaimed Fidelio. 'Old Chico has told me of him.'

'Well,' continued Don Victor, 'Moctezuma lived in a huge stone palace with many, many rooms. His servants were more than he could count. Now, he had a brother, who loved flowers and birds greatly. He built a great garden at a place called Itzapalapan, where he kept every bird in the kingdom.'

'Every bird?' queried Fidelio, his eyes opening in astonishment.

'He had over three hundred servants whose only duty was to look after the birds and feed them.'

'Just like me,' said Fidelio, remembering this was his task.

'Yes, just like you,' Don Victor went on. 'In that aviary he had the *cetzontli* which sings so beautifully. He had the little *hilguero*, which you always try to catch, and cannot. He had fine parrots and turkeys, white egrets and—in fact every bird that the Indians would bring him. These servants, just like you, would look for the favourite food for the birds and many times it would have to come from many days' journey, for the cage of the birds at Itzapalapan was high up in the *montaña*, higher than Portillo Grande here; in fact it was twice as high.'

'You said they had every bird, Don Victor. Did they have the quetzal?'

'No, they did not. That is just what I was going to tell you. Although the Indians were very skilled in catching every bird that lived in the jungles, near the sea-shore, or in the pine forest—oh, yes, they captured —they could do that, but they could never keep alive the quetzal bird.'

'Not at all? Why, we have kept them alive!' said Fidelio, engrossed in the tale.

'You see, you and I have done what even the great Lord Moctezuma could not do. He could command an Indian to die, and the Indian would kill himself. He could say, "Bring me fresh fish from the sea," and the Indians, by relay, would bring fresh fish from the sea. But when he said, "I want the quetzal kept alive, so that I can see this beautiful sacred bird alive," that was one thing he could not command. And we have done it! Think! Fidelio, we have done it! At least so far.'

At last Fidelio understood. The sacred bird, quetzal, never kept alive in all the history of the great Aztecs, never by anyone else in the whole world . . . and they were succeeding. Now he must protect them more than ever. They must get the birds out alive.

Then he said, 'What did they do with all the birds in the cage at Itza—Itzaclapon?'

'At Itzapalapan,' corrected Don Victor. 'Why, they kept them in the cage for the chieftains to look at. When the birds moulted—you know, when the old feathers dropped out and new ones came in, the Indians would go into the cages and pick them up.

These would be brought to the weavers of birds' feathers. They were feather artists, and with the red, blue, green, brown, white, and black feathers moulted by the birds the royal weavers would make beautiful embroideries, which they sewed on the tunics of the Aztec lords. Then on great Fair Days the Lords of the Aztecs would wear them.'

'What about the long green feathers of the quetzal?' burst out Fidelio, unable to keep silent any longer.

'Well, these feathers could be worn only by the chieftains. If any other Indian was seen with them, he would be killed by the chief's soldiers. Why, the feathers of the quetzal were valued more than gold.'

'More than gold?' repeated Fidelio, astonished.

'Yes, more than gold. Moctezuma was afraid, if the nobles continued to want the quetzal plumes and the birds could not be kept alive in the great aviary, that there would soon be none left in the forests.'

'Ai, for the birds would all be dead,' murmured Fidelio.

'Moctezuma made laws to protect them. No Indian was allowed to kill a quetzal bird. In order to get its plumes, the bird must be caught alive, the six long tail feathers pulled out, and the quetzal released unharmed. Anyone who killed a quetzal was punished by death.'

'Maybe that is why we Indians,' said Fidelio, 'are always afraid of something when we kill a quetzal bird.'

'Now there is a great puzzle,' Don Victor went on, 'about this. How did the Indians catch the quetzal alive to get its feathers? I once read a book by an old

Spaniard named Hernandez. He lived in the time when the Indians were great. He said they would go into the jungle, like ours here, where the quetzal lives, and put sticks into the ground, covered with bird lime.'

'Bird lime,' repeated Fidelio, 'what is that?'

'It is a sort of sticky stuff, much like what comes out of a rubber tree when you cut it with a knife. About these sticky posts the Indians would put ground corn, just like the balls you make for the quetzalitos. Hernandez said the birds, seeing the corn, would fly down to the ground to eat it and get stuck on the sticky posts. What do you think of that story?'

'I do not know—it is strange.'

'You and I both know the quetzal with its beautiful tail would never come down to the ground to eat the corn.'

'That is true, señor. Our quetzal is too proud to drag his tail in the earth,' agreed Fidelio.

'So that story is not true. But how did they get the birds alive?'

Fidelio put his head between his hands and stared hard at the ground. He was thinking!

'I can hear the wheels turning, Fidelio! What is it you are thinking?' asked Don Victor after a few moments.

Fidelio joined him in laughing, although he did not quite understand. At last he went to his little room, for he now stayed with the white people. He brought back his blowgun and numerous little clay marbles, just the kind Miguel used during the first days in the forests. He put these in Don Victor's hands.

'And what is this for?' asked Don Victor.

'This is what I have been thinking about,' said Fidelio.

'And——' Don Victor went on, curiously.

'With this blowgun an Indian can bring down a quetzal without killing him. I can't. I am still too small and can't blow hard enough, but Miguel can. I'll talk to him.' Fidelio ran down the hill to the house of Miguel.

Presently Miguel returned with Fidelio and confirmed what Fidelio had said. Yes, he could 'shoot' a quetzal and catch him alive.

'Seeing is believing,' doubted Don Victor, and sent him to the jungle after the quetzals to see if it were true.

'Why do you take the quetzalitos away from here, Don Victor?' asked Fidelio, presently.

'In the land far away, we have an aviary as great as the one belonging to Moctezuma's brother. We too have many birds from many lands. But we have no quetzal. I wish to bring back a quetzal that people there may see how beautiful they are, and know of their fine green tail feathers.'

'Fidelio,' Doña Cristina called the Indian boy. 'Come, Fidelio, it is time for the quetzalitos' cod liver oil.' Fidelio ran quickly, for he liked the white woman very much. He held a small teaspoon while Doña Cristina poured a few drops of yellow oily liquid into the spoon, which had a few drops of water in it. He dropped the mosquito netting of the cage and the quetzalitos hopped near. He held the spoon to their mouths and they opened them and gulped at the liquid. It went down, but they did not like it. They shook their heads and rubbed their beaks on the side of the perch, and jumped away again.

'Why must we give them this "cood leefer ool"?' said Fidelio. He could not pronounce the funny words correctly.

'It makes them strong and healthy,' said Doña Cristina.

'But the quetzals in the forest are strong, and they do not get "cood leefer ool".'

'That's just it. They live in the forest; eat the things they want; then fly for exercise. They are not kept in cages like these birds. Do you remember the quetzal that Beltran brought us?'

'You mean the one with the crooked legs, that could not stand up?'

'Yes, Beltran tried to raise that bird. He kept it in a cage, never exercised it, and gave it no cod liver oil. Because of this, the bird's legs never became strong. Now, run along, you haven't found the worms for the birds yet today.'

Fidelio shouldered the heavy *pujante* and went to the forest to dig in the earth for worms and beetles. He turned over old logs, finding squirming white larva under the cool, moist places.

'And,' he said aloud to himself, 'Moctezuma had many birds, more than I could count. Did they all have to have so many worms and exercise and "cood leefer ool"? It must have been a lot of work.'

In the afternoon, Miguel came back bearing in his hand a crude basket made of vines; he held it loosely. Fidelio knew what he had in it, for he saw the long

green tail feathers which Miguel had let hang outside.

'Is it alive?' asked Fidelio. Miguel nodded his head.

Don Victor took the bird into his tent and, closing
the flap, laid the package on his bed. There in front of
his eyes was a live male quetzal. He was somewhat
stunned, and could not stand up; he had been so long
in the basket. Presently, with dignity and hauteur, not

even deigning to look at his captors, he sat up on the
edge of the bed, swaying slightly. Then he sprang for
the window of the tent, thinking it was open air. He
fell back, and settled in the corner, his long tail lying
on the bed.

'Now, tell me how you did it,' Don Victor asked
Miguel, for the Indian was standing there leaning on
his blowgun.

'I go into the forest and make a noise like the she
bird—like this—"Whe-ooooo, whe-ooooo", under an
*aguacatilla* tree. Well, at last comes the big quetzal.

He says, "Whe-oooo, whe-oooo"; then I say much louder, "Whe-ooooo, whe-ooooo". He comes down and I shoot. Not hard. The blowgun ball hits him in the head. And ump—he flies down. So I catch him.'

'And you can do this often, any time you wish, and always get them alive?' demanded Don Victor.

'Usually. Sometimes the ball hits hard, and the *ampusay* dies. But not often.' And Miguel left, happy with the money Don Victor gave him.

'Cristina,' called Don Victor, 'come and see the great male quetzal Miguel brought.' When she had arrived, with Fidelio beside her, 'I think, Cristina,' he continued, 'that we have solved another puzzle in the quetzal legend. This may have been the way the Old Indians got the bird alive.'

He went over to pick the quetzal up, but the bird would not let him, flopping about until at last he was put into a cage, where he sat quietly. He would not eat. He would not drink. They had to force him. This much, then, of the quetzal legend was true. Neither males nor females, when full grown, could become accustomed to captivity, and they would soon pine away for the great cloud forest which was their home.

7

# THE ARMY ANT INVADES THE CAMP

Two months had passed since the white people had come, two months full of everyday events but without much excitement. It was now September. The quetzalitos had grown until most of them were a solid green. Their crests began to form. Red feathers appeared on the breasts of the males, grey on the females. The routine of feeding went on with deadly monotony.

The Indians would come sometimes to see the birds. They could not understand how the white hunter could do this wonderful thing. Now and again one would bring an adult bird to the camp, hoping for an extra piece of silver. Don Victor, always hopeful, would try different ways of adapting each one to captivity, but never kept them for long. They always wilted and refused to eat, and they had to be set free.

Most of the Indian boys were jealous of Fidelio, for he had new clothes and a new hat, of which he was inordinately proud. He would admire himself in Doña Cristina's mirror until Don Victor told him he must think he was a papa quetzal! Fidelio would grin, and go back to his birds. The boys were more direct and less kind.

'Fidelio . . . he is a bird himself! He is a mamma quetzal. He hops about to get food for his little birds. He looks for worms. Just like a bird!' taunted Ignacio one day, hopping about in imitation of Fidelio trying to find grubs for his charges. The boys standing about laughed and added to the remarks.

Fidelio did not bother to answer them. These stupid boys wanted to be always just the same, Indians—to sit on their haunches, grow corn, and eat nothing but tortillas and beans all their life. Ever since Old Chico told them about the white men he had wanted to see other things. Let the boys tease him! He liked to be the 'quetzals' mamma', even if they did eat too much to please him.

These silly boys cared nothing for the stories of the Plumed Serpent except as stories around the fire! Even if Fidelio had foolishly thought Don Victor was 'the fair god' at first, he knew there were many things to see. For hadn't Old Chico told him that his father's father's father's father's father collected tribute for the Mexican kings at Naco? Let them laugh! He, Fidelio, knew!

Even the older Indians were sullen, and prone to play small tricks on Fidelio, as Don Victor gave them no more money, there being nothing more for them to do. The birds had been caught. There were twelve fat little quetzals, and that was enough.

'We must get ready to take the birds out soon,' said Don Victor one day.

'To the coast, where the big ships are?' asked Fidelio.

'Yes, I will take you, but first I must ride down to see what the conditions are,' was the answer.

'Con-di-shuns—what are they? A house for the birds to live in?' inquired the ever-questioning Fidelio.

'Conditions, what the road is like, how hot it is, and who will meet us—and also to see the state of the river,' answered Don Victor. 'Now, Fidelio, this is the most difficult part of the whole undertaking. Here we are near the forest, and it is high. There, it is low, and hot!'

'The poor little birds, they will not like it hot. They like the cool winds and the clouds. It is true, it will be hard. But do not despair, Don Victor. I, Fidelio, will guard the birds.'

Don Victor had the mules saddled, took his gun, for it was said that bandits lurked in the road; put food in a pack, for houses were far apart; and took his rubber poncho, for it rained often now, and the rain would increase as October approached.

Before he left, Don Victor talked quite seriously to Fidelio. 'I am counting on you,' he said, 'to help Doña Cristina take care of the birds and to mind her. You must take care of her, too, for she will be alone except for you. Guard her well. I am leaving you with a gun. You see, a great deal will depend on you.'

'I will watch, señor, truly,' promised Fidelio, fidgeting a little at Don Victor's seriousness.

'You know that in my language Fidelio means fidelity, or a person who is loyal, faithful, and true. Remember, the success of the whole trip may depend on you.' Saying good-bye to his wife, Don Victor mounted his mule, and rode down the mountainside.

He had been gone four days before anything beyond the usual régime occurred. Fidelio carried out the same routine as he had when Don Victor had been there. At six each night the quetzalitos were carefully placed in smaller cages, where nothing could get at them, and these were put in a small house next to Fidelio's little palm-thatched hut.

Often he would look longingly at the gun and think how he would use it to protect Doña Cristina. He was a little disappointed that nothing had happened to disturb their peace. 'How will Don Victor know how faithful I am,' he asked Doña Cristina, 'if I get no chance to shoot the gun?'

Doña Cristina was not as anxious for trouble. 'Oh, Fidelio, I just hope you won't need to shoot the gun. I like it quiet and dull.'

Fidelio's wish for excitement was soon fulfilled. He went to bed one evening and, as always, fell into deep sleep. How long he had been asleep or what waked him he could not tell, but suddenly he was wide awake. His primitive Indian instinct told him something was wrong. There was a rustling sound in his room. He sat up and listened.

Reaching for a stick of pine wood he kept near him for a torch, he felt for a match in his pockets. Before he had lighted it he thought, 'What if it is the Sisi-miki?' A lump came in the pit of his stomach, and fear was all about him. The dark seemed alive and moving. He was afraid to make a light. In the other room he heard the birds chirping restlessly, until he really became panicky. He sat there frozen, not knowing what to do. As he clutched the pine torch tightly to him, his brain was convulsed with fear. Then he thought of what Don Victor had said. This was the chance he was wanting. Moving his hands quickly, but still somewhat stiffly, he lit the piece of pine. For a moment he was blinded by the light, then he rose from his bed, putting his feet on the floor.

'Ouuu-ch——!' He jumped back and rubbed his feet. What was that? Fire? He peeked over the edge of his bed, holding his torch to the floor. Why, the room was alive with great black ants! Black legions of the worst enemy in the jungles, the army ant. '*Tumecas! Tumecas!*' he cried.

They were crawling up the bed now, and Fidelio could not even get his sandals. He heard the birds chirping louder. Gauging the distance to the door, he made one leap. Ants clung to him, biting and stinging, as he gained the doorway and struck them off.

Running to the tent of Doña Cristina, he shouted hysterically, '*Tumecas!* Doña Cristina, hurry, hurry. It is the *tumecas*, eating the quetzalitos.'

Doña Cristina came out barefooted, but seeing the ants she rushed back to put on her boots. 'Quick, Fidelio, get a bucket of water.' She routed out the servant-girl, and Fidelio ran for the water as if the Sisimiki were truly after him. Running back with the water spilling over him, he dashed to the birds.

'Throw it on the floor beside the birds' cages,'

commanded Doña Cristina, grasping a broom and following close behind him. He poured it all over the

ants; rushing back for more he threw it over the birds' cages. The ants floated away, and struggled to escape from the flood. Doña Cristina swept them away from the cages of the birds as fast as she could. Still they came!

'Fidelio, get the cages and carry them out of the house,' she called. He did as she said, opening the door and taking out the birds which were already covered with the ants.

The servant-girl, by this time, had a fire going and was making coffee for Doña Cristina, who with Fidelio was busy with the birds. Up from the village came a straggling of curious Indians, drawn by the noise of Fidelio's shouts. Instead of helping or calming the frightened birds, they stood around laughing heartily at the unwonted antics of Fidelio and his mistress. They not only laughed, but among themselves they predicted worse havoc. It was against nature to keep the *ampusay* alive in cages. The ants were a judgement —perhaps even the Sisimiki had—at the muttered word the Indians stopped laughing, and began uneasily moving back to their huts. It was dangerous to be out at night like this—they might be hurt.

Doña Cristina was secretly quite relieved when the natives left. She gave Fidelio a cup of coffee, and as he began to blow on the hot brew, she praised him.

'If you had not wakened,' she said, 'the birds would have been eaten alive.'

'I thought it was——' Fidelio suddenly looked about him at the moon high overhead, the deep shadows and eerie light of the night, and stopped.

'I know,' she answered, 'but you were brave. Don Victor will be proud when I tell him.'

They brought the cages back. All around were remains of the struggle. The ants had overrun the house, eating everything possible, including cockroaches, which Doña Cristina had tried in vain to eliminate, beetles, and anything they could find.

Doña Cristina shuddered as she realized what a close call it had been. These creatures marched in seemingly endless columns and nothing stopped them for long. They went on and on! They were more dreaded than wild beasts. There was something horrible about their blind savagery.

When Fidelio and Doña Cristina got all the ants from the birds' cages, the quetzalitos were brought into her tent. But they were fully awake and restless, anxious to jump and fly. It was almost morning before the birds fully recovered from their fright, and Doña Cristina and Fidelio felt that it was safe for them to be left alone.

# 8

# THE SISIMIKI LEAVES ITS MARK

It rained and rained. The skies opened up and poured down heavy sheets of water. Small streams that were only rivulets rose until they were rushing rivers. The quetzals seemed now to be aware of all this. They would stretch their wings, first the left one and then the right, almost hard enough to make them fall off their perches. When Fidelio sprinkled water on them, they moved about, shaking their feathers as if they were getting a bath. One stormy day, when the rain was coming down very hard, one of the quetzalitos burst into its song of 'whe-oooooo, whe-oooooo', just like the big birds. The sound of its own voice startled it, and it looked about as if the noise had come from another bird.

'Look, Doña Cristina,' exulted Fidelio, 'he is speaking in grown-up language, our quetzalito. Is he not clever?'

'I wish Don Victor would return. The rains make me afraid, and it is so muddy. I think I shall die if I don't get out of all this mud.' Doña Cristina was beginning to be nervous. Don Victor should be back at any time.

The rainy days had passed, and the sun shone again. Everyone cheered up, in spite of the fact that they all knew there would only be a short interval before the *chubascos* began, the hurricanes from the east which would cause the sky to get yellow and then black. Trees would be uprooted, houses torn down. The Indians feared these times as well as the birds, which had to be grown and out of the nests before the winds began. The old dead trees used for nests would come crashing down. By the last of September all the small quetzals were ready to fly.

The restlessness of the Indians disturbed Doña Cristina more than she realized. Even Fidelio, who was used to them, noticed that they were less friendly and more inclined to be sulky.

'Chico says they should not have helped you get the birds, señora,' he confided one day; 'he says perhaps the Sisimiki will take its revenge for meddling.'

'Nonsense,' she replied, but she kept close watch for her husband just the same.

It did not help matters when Old Chico's pig died. Now that seems to have had nothing to do with the quetzal birds or Don Victor, and perhaps it had not, but Indians think differently from white people. They believe in enchantments. They believe that people can put spells upon other Indians and their animals. They were still very much upset that Don Victor had persuaded them to go up into the *montaña*—the very lair of the Sisimiki. True, nothing had happened so far! But, yes—had not the *tumecas* descended upon the white man's camp? Was not that a sign? Even though nothing had happened to themselves, they all feared the worst.

When Old Chico's pig died, the temper of the people flared up, and fear stalked the village. For about the pen of the pig were the footprints of the Sisimiki! As plain as the footprints that Fidelio had seen one day in the woods! There was no denying that the pig was dead—he lay cold and stiff before their very eyes.

The white man was the cause, beyond doubt. He had made them go up into the *montaña*. They had tried to dissuade him. Fidelio, too, he was to blame. They should not have been talked into going up into that *montaña* to get the quetzal birds.

The Indians went up to the camp, and brought Fidelio down to see the marks of the Sisimiki, and he had to confess that they were the same marks as those he had seen up in the *montaña*. They wanted to go at once to destroy the birds.

Fidelio was frightened. He was frightened in more ways than one—the Sisimiki was dangerous, and fearful—his heart pounded in terror at the thought. But, also, he was afraid for his quetzalitos, his little birdlings that he had loved and cared for all summer. He was afraid for his friend, Doña Cristina. He did not know what to do, or where to turn.

Perhaps, after all—perhaps the Sisimiki did not want the birds to leave the *montaña*! He did not know. He did know that he wanted his quetzalitos left alone.

'No, no,' he shouted, 'not the little birds! The quetzalitos could not have anything to do with it.'

'Maybe the Sisimiki is coming to look for them. Perhaps if we let them go, it wouldn't come down here again. Do you want us to suffer more than we have?' the others insisted.

Old Chico, too, had changed. He had never forgotten the days of the tribal witch-doctors, even if they had been baptized by the *padre* and prayed to the Virgin. He still believed deeply in enchantments. 'Yes, I think they are right. We must let the birds go. You, Fidelio, must not work for the gringo any more,

D

or you will become enchanted like my poor pig.' A big tear rolled down his cheek, down to the long hairs on his chin, and rolled down the entire length of the chin whiskers to drop on his shirt.

Fidelio was thunderstruck. He could not leave the birds now. Don Victor depended on him. He would arrive any time—perhaps today. If he could only keep them quiet until then.

'All right, Old Chico, I shall do as you say. But I must wait for my wages. The white man owes me ten pieces of silver. If I left now, I never would get it. I have worked hard, don't you think I should get my pay?'

Some of the Indians, who were greedy, and knew the little boy would not keep all this silver to himself, agreed that he should stay until the white man came. The Indian women, though, were hysterical with fear, for they thought if the Sisimiki came for the pig, the next thing it would want would be their children. Next it would come for themselves, and they wailed in their houses. They cursed Fidelio, as he left to go back to the camp, for bringing this evil upon them.

Fidelio did not tell Doña Cristina about the Indians' desire to let the birds go, for he thought they might change their minds. Besides, he hoped Don Victor would return—for he would settle everything. Alas, for his hopes! Early next morning the Indians all appeared and demanded that the birds be set free. Fidelio, trembling but determined, refused.

'Doña Cristina,' he called, and when she appeared, he spoke. 'The Indians say the Sisimiki is angry—they want you to let the birds go.'

Her answer was to go for the gun, but it wasn't there! Fidelio had taken it the night before, and it was in his room.

Seeing a big male quetzal, the last one Don Victor had tried to tame, the Indians moved to his cage and knocked it over. The door opened, and the bird took flight. The Indians not only looked pleased, but they shouted with glee! They would soon exorcise this spell. They approached the other cages. 'Now for the little birds,' and the boys ran after Fidelio, holding him, to keep him from getting the gun. The first cage was opened and two quetzalitos were released, when a shot rang out!

At this sudden interruption, the Indians looked about, frightened. Don Victor rode in, got down from his mule, very angry, saying, 'What's all this about? Why this disturbance?'

The Indians saw the gun in his hand and knew he was angry. Santiago said, 'The Sisimiki killed Old Chico's pig last night—because we went up into the *montaña*. It comes for the *ampusay*. If we let them go, it will not come again. We Indians are afraid. The Sisimiki may come for us!'

'You're crazy, all of you. Your Sisimiki doesn't exist. Go now, all of you. I will come down and see what is the matter with your pig, and why he died.' Don Victor spoke severely, and the Indians, cowed, like bad children, went away without saying a word.

Fidelio rushed up, crying, 'They let out the big quetzal and two quetzalitos. We could not stop them. We tried, Don Victor, we tried.'

'Never mind about the birds,' said Don Victor. 'I

was going to let the big one go anyway. And we have ten quetzalitos left. Don't worry! It's all right. I can't get too angry at the fools. I need them to help me carry the birds out. I have to have two men. I can't be too harsh with them. After all, they're like children.'

'But they won't go with you,' protested Fidelio, torn between relief at Don Victor's return and fear of the Sisimiki and its enchantments. 'Even Old Chico said I must stop working for you.'

'And all because of the Sisimiki?'

Fidelio nodded his head.

'We'll see what can be done,' and Don Victor went into the tent to talk with Doña Cristina.

When he came out he said, 'Come, Fidelio, let us go down to see Old Chico.'

# 9

# THE WHITE MAN DISPELS THE LEGEND

Old Chico sat by his house, still sad and gloomy about his old pig. He was not very pleasant when Don Victor and Fidelio came up.

'I am sorry about your pig,' began Don Victor.

'It is your fault that the pig died,' claimed Old Chico.

'How could I be to blame? . . . I was not here.'

'The Sisimiki is angry that we helped you get the birds. It has killed my pig. The tracks are there.' He waved his hand towards the pig pen. 'Until we let the birds go, everyone in Portillo Grande is in danger. Who knows where it will strike next? We should never have gone up into the *montaña*.'

'Let me see these footprints,' requested Don Victor, amiably, trying not to antagonize the old man any more.

Old Chico took him to the sty and showed the footprints, which were small and somewhat resembled those of a child. 'That's it,' he cackled excitedly, 'those are the marks of the Sisimiki!'

Don Victor got down on his knees on the ground and examined the marks very carefully, even using a magnifying glass to impress Old Chico.

'So,' he remarked, 'you think the Sisimiki killed your pig. What is it like, this creature you are afraid of?'

'It is a beast, señor, most horrible,' Old Chico declared. 'It is about as big as a little child, and is covered all over with hair. It looks like a monkey, but it walks upright on the ground like a child, and kills babies and other people. It puts spells on us who offend. It is very dangerous even to speak of it.' He looked fearfully over his shoulder as if to see the creature materialize out of thin air.

The Indians, who, full of curiosity, had gathered around the two men and Fidelio, stood about saying, 'Yes, señor, it is true what he speaks. Our children will all be killed, and our crops eaten, and we shall starve.'

'Well,' declared Don Victor, 'I am going to show you something. I shall fix your Sisimiki so it will never bother you again.' He left with Fidelio trailing at his heels.

'We have but one way out,' Don Victor told Doña Cristina and Fidelio when they were back at the camp. 'I must find this Sisimiki. If it is what I think it is, and we bring it to them, the Indians will see that this mark is not made by their imaginary Sisimiki.' He turned to Fidelio. 'Have you ever seen the Sisimiki?'

'No, señor, I have but seen the tracks.'

'Do you know anyone who has seen the Sisimiki?' Don Victor continued.

'No, señor, none of us has seen it—but we know what it does. No one is quick enough to see the Sisimiki. It comes, it goes—we do not know.' Fidelio, still afraid, felt he was not making out a very good case for the Sisimiki. How could he make the white man know how horrible the thing was, how silently and dangerously it came and went?

Doña Cristina was not enthusiastic about Don Victor's plan.

'This is the only thing we can do, Cristina,' Don Victor insisted. 'I must have men to carry the birds out. The roads are awful. We have to cross the Cuyamapa River five times. It will take four days to make the trip, and we must leave no later than next week, before the *chubasco* comes, or we shall never get the birds out alive.'

'What will you do, señor?' asked Fidelio.

'I am going to the woods tonight and, if I am right, I shall bring back that monster with me to-morrow.'

Fidelio was astounded. His eyes grew round with fear and his mouth opened, letting his voice out hoarsely. 'You, Don Victor, you are going to bring back the Sisimiki?'

'Yes,' said Don Victor, 'I am going to bring back what you think is the Sisimiki—and alive.'

That afternoon he left for the *montaña*, with a gun and flashlight. He carried a little sack and a large net. He did not go off on the road to the *montaña*, but by way of the cornfields.

'Would you like to come with me, Fidelio, to see me catch the Sisimiki?' he asked. Fidelio hid his face, and looked down to the ground. Nothing would have dragged him up the hills after a Sisimiki.

All afternoon he kept fearful watch for his friend and master. Dinner-time came and went, but no sign of Don Victor. Doña Cristina was not too cheerful herself, but Fidelio was in an agony of fear. He knew how strong Don Victor was, but the Sisimiki was the unknown—the mysterious and unpredictable creature of spells and enchantments over which men had no control. He truly believed that Don Victor would not return, or would come back in terrible agony.

Late that night, he heard someone coming. He had drifted off into an uneasy slumber, pierced by miserable dreams, but awakened at once at the sound of footsteps approaching. He lit his pine torch and held

it high. He was astonished to see Don Victor return so early and safely. He looked his master over carefully for the wounds he would surely bear. On his back, Don Victor carried something in the net he had taken. It seemed rather heavy. He handed it to Fidelio.

'Well,' he laughed, 'inside is your Sisimiki.'

Fidelio dropped the net with its burden inside, and ran.

'Are you like all the rest, afraid of your own shadow? Here, put it in the kitchen,' Don Victor commanded.

'Tomorrow we shall see something.' But Fidelio had disappeared. Doña Cristina, at the sound of the returned hunter, came out to greet him and took the net.

Very early in the morning, Don Victor with his burden went down to the village. Fidelio, afraid of what was in the net, kept far behind. Old Chico saw them coming, and called all the other Indians, who were just about ready to set off for their *milpas*.

Don Victor held on to the net. 'Tell me, Old Chico, do you know the footprint of the Sisimiki?'

Old Chico nodded.

'And do all of you know it?' Don Victor swept his arm towards all the other Indians.

They all nodded and murmured, '*Si*, señor, we know the mark.'

'And,' Don Victor went on, 'do you say this is the footprint of the Sisimiki?' He pointed to the prints in the mud beside the pig pen. Again they all nodded and agreed he spoke the truth.

There was a tenseness in the group that made Fidelio afraid. How could Don Victor win them over? He seemed unworried, though, and spoke quietly, as always.

'Now you all think hard. Is there any other animal that makes a mark like this? Only the Sisimiki?'

'Oh, no,' they shouted in chorus, 'no animal in the whole forest makes a mark like that. The mark of the Sisimiki is its very own and this is it. There is no doubt.'

'And it came down to kill the pig,' shouted one, growing increasingly excited.

'Soon we will all be enchanted,' wailed one of the women.

Don Victor spoke more loudly. 'Silence, all of you, and see what I have brought.'

His insistence stilled them momentarily, and they watched fearfully as he untied the net. Fidelio stood far off, ready to run.

The mouth of the net opened, and instead of the great monster, a horrible half-human creature, out came a small furred animal. Walking on all fours and looking something like a raccoon, it carried its ringed tail held high over its back.

'Why,' gasped Fidelio in astonishment, 'it is a *tejon*!'

The little animal walked across the soft mud beside Chico's pig pen, near the tracks left by what the Indians claimed was the Sisimiki.

'Now, come here, Old Chico,' said Don Victor. 'I want you to examine these footprints. Look at those you say were made by the Sisimiki, and then at these of the *tejon*.'

Chico and the others did so. They fell back utterly confused and astounded. The tracks were the same. Those of the *tejon* and those of the Sisimiki were identical.

# 10

# THE TREK BEGINS

In four days the trip to the coast would begin. Cages were being made, food prepared, and boxes put in readiness. It had rained and rained until the foot-roads were like small rivers. Don Victor looked anxiously at the sky and hoped the full moon would stop the rain for a while. They would have to ford the rivers, and they never could in this rain, for the rivers would be too high.

Old Chico came with his sons to ask if they could be of any help. He was very sorry now for what he and the Indians had done. When they learned that the creature they had been afraid of all their lives was nothing but a *tejon*, or coatimundi as some called it, a little animal whose only fault was that he ate their corn, they saw how recklessly they had acted. They remembered how the *padre* told them the Virgin would protect them, and they had forgotten. They were sorry they had let the birds out. Don Victor did not tell them he had intended to let the big one go anyway, for he knew it could not survive the trip out. As for the two little ones, he still had ten left.

The Indians agreed to carry the quetzals out on their backs, now that they no longer feared the Sisimiki. True, there were still some Indians who insisted that there was such a horrible animal in the *montaña*, yet since Don Victor had proved at least that the footprints they saw were not made by this creature, the Indians said nothing more, publicly.

Once more Don Victor took Old Chico into his confidence:

'I want to leave at one o'clock in the morning, in the full of the moon. The men will walk with the bird-cages strapped to their shoulders. Doña Cristina and I will ride. We shall have to move quickly, because after we leave here, you know, the hills become low, and it is hot down there.'

'*Si, si*, señor,' replied Old Chico, 'I know that. But I am very much afraid you cannot get across the river. The Cuyamapa is very *bravo*, very wild.'

'That is the difficult part. However, we cannot wait, Chico. This is the beginning of October. It has stopped raining for a few days now. You know after this dry spell what will happen. The *chubascos* will come—and we could never get out.'

Chico told his sons they must be ready to go in four days. Fidelio went up to Chico. 'Old Chico, Don Victor said I could go to the coast to see the ships. He promised me a long time ago, that if I helped raise the quetzalitos I could go. You will let me, won't you?'

'I do not know,' answered Old Chico; 'since your

father died, I have been your *padrino*. You know I would be responsible if anything happened to you.' He looked down rather fondly upon the upturned, eager face of Fidelio.

'What could happen to me but good?' asked Fidelio. 'Would I not be with Don Victor?'

'I expect you will be safe with the white man. Yes, yes, you can go.' Old Chico went down the hill, still pulling his long chin whiskers between his fingers.

The last days were full of suspense. The quetzalitos were big and strong. All of them had grown so large that Fidelio could hardly tell them apart, for since growing up they had lost much of their character. One used to be fat, another thin. Now they were all equally plump. The little birds were trained to run on the floor during exercise-time just like chickens. Fidelio could see the beginnings of the green tails. Each bird had two green feathers, a little longer than the black and white feathers on the tail. Don Victor used to join Fidelio every morning after he had given them their 'cood leefer ool', and together they would inspect the feathers of the birds to try to tell the males from the females. Two of them were certainly males, for their breasts were already covered with red feathers.

Fidelio had grown to love the little birds as much as the white people did. Every night when he put them in their cages, he would talk to them. Not too loud, for he did not want anyone to hear. He had a name for each one, and every time he put it in the box he would say, 'Now, fatty quetzalito, I have fed you and brought you big, fat worms, and soft, good avocado. I have taught you to run on the ground and to hop on my finger. Every day, whether you liked it or not, I gave you your "cood leefer ool" to keep your legs from growing crooked. Now I ask you one favour. Please don't die when we carry you to the coast.' The birds would give him a look which made him sure they understood.

Now came the final night. The tents of the white people were taken down, and packed into big sacks. All the boxes were closed and put on the mules. One of the Indians started off in the afternoon to get far ahead of the others who did not go until night.

Don Victor and Doña Cristina went down to the village to say good-bye to the Indians and to thank them. They brought each a present, a knife or pipe for the men, cloth or kitchen pots for the women. Fidelio thought they should be sad that the white people were going, but the Indians did not seem to care much. Fidelio was glad he was not left behind, but was a part of the bustle and excitement of the approaching departure. Adventure lay ahead! And ships that carried fire in their middle! It was glorious!

That evening he put the quetzalitos not into their regular sleeping cages but into the travelling boxes. Stronger than the ordinary ones, they were covered like the rest with mosquito netting, and had straps for the Indians to slip across their shoulders. The quetzalitos were fed until they were too full of food even to jump. Placed in their boxes, they were ready.

Everyone went to sleep until the full moon rose.

It was one o'clock in the morning, and cold when the moon rose over the mountains—immense, bright, full! This was the signal. A stir was felt throughout the camp and all rose silently. Don Victor and Doña Cristina mounted their mules, the Indians shouldered the cages of birds, and the straggling little group passed quietly down the mountainside. Old Chico got up to see them go. He watched them as long as the light of the moon showed the moving figures, and then the party passed into the great pine forests, and Old Chico saw them no more. He returned to his hut, somewhat sad, for they had carried away the symbols of the Plumed Serpent, and what was left of his people's greatness.

All night long the little party walked. It was still and cold in the forest. The Indian walking ahead carried Don Victor's flashlight to point the way. Everyone was silent and thinking the same thing. Would the birds live?

At dawn they came out of the pine forests. The rising sun touched the tops of the pines until they became red enough to make Fidelio, looking up, think they were on fire. From the pines hung moss, long and grey, swaying gently in the morning breeze, and looking like the long grey beard of the Spanish *padre* who had come to baptize this little Indian and call him Fidelio.

The birds were silent, and too frightened by the movement of being carried even to get down from their perches. At last the Indians put down the cages, and ate the cold tortillas they had brought along for their food. Don Victor and Doña Cristina shared theirs with Fidelio. All were tired. No one spoke. The birds were fed, and the party went on. Fidelio did not find this part of the trip very thrilling.

Fidelio, standing on the rim of a great hill, could see a river below, far below, wide and very swift-flowing. He asked the Indian in front of him what river it was, and was told, 'Cuyamapa'. It was the one Old Chico was afraid of, that was very *bravo*, very wild.

Then they came to the edge of the hill, they went down, down, going round and round in semicircles, for it was too steep for them to go straight down. When they reached the base it suddenly became very hot. As the birds were used to being in a cool atmosphere, they grew restless.

'We shall have to stop here,' said Don Victor when they entered a patch of forest near the river. 'It is noontime now, and too hot to go on. We shall sleep now and tonight start off again. Fidelio put up the mosquito screen and take out the birds. See that they are fed, and then you try, too, to get some sleep.'

Doña Cristina was already asleep on the ground. They rested all the day until the moon was high again, for in the low altitudes of the tropics it is very hot during the day.

Again they walked, crossing the Cuyamapa many times. At first the river was shallow, coming up only to the Indians' waists. When it was too deep, they carried the bird-cages on their heads.

Later, as the river grew wider and deeper, they had to build rafts. The Indians felled eight large balsa-trees, a light wood that floated easily, and tied them together with vines, driving pegs into the soft balsa-wood to secure them.

The Indians put the cages on these small rafts, and

swam beside them across the river. Their clothes were wet and in the depths of the night it was very cold. The Indians complained, but not Fidelio. In spite of the drawbacks and hardships, he felt he was taking part in something new which had never been done before. He was happy that he, just another little Indian, could help to bring the quetzals alive to the great aviaries in the far country.

The sun rose hot and fierce. 'The day,' said Don Victor to Doña Cristina, 'will be a scorcher.'

Fidelio wondered what a 'scorcher' was, but he was too tired to ask. There was no shade here, as it was all open country. Don Victor said they could not rest until they got to the town of Morazon.

'We should be there by noon,' he added. At ten o'clock, Fidelio felt he could not go on, the sun beat down with such fierceness. He was, after all, only a little boy and not as strong as the other older Indians. Still he plodded on, his feet hurting him until he was limping badly.

Don Victor noticed and took pity on him. 'Fidelio,'

he shouted, 'tell the Indians to stop. Go cut some banana leaves and put them over the bird-cages to shade them.'

When Fidelio had done this, Don Victor told him to mount his mule, for he wanted to walk. Fidelio was afraid, for he had never been on a mule, but Don Victor insisted. It was funny to go along on muleback. As Fidelio swayed to and fro, he remembered vaguely that this was much the same sensation he used to have when he was a little, tiny boy. His mother used to strap him to her back while she worked in the fields, and now he was swaying back and forth again. He could see into the cages of the birds, too, and not one of them was on its perch. All were down, huddled miserably in the corners of the cages. Don Victor noticed it also, and frowned all the while.

# THE TREK ENDS

After two long hours that seemed longer than both the days they had travelled, they came to Morazon, which was just a score of whitewashed buildings grouped about a big white church. Don Victor directed them to a house, where he had previously arranged they should stay.

Wearily the Indians put down their burdens, and fell down, going right off to sleep. But there was no sleep for the white people or Fidelio. The birds must be cared for at once. Fidelio reached into the boxes and toppled the quetzalitos out.

How pitiful they were! Their feathers were all untidy. Most of the birds could not stand. All were dizzy. Many looked as if they were dead. Don Victor cursed. Fidelio picked up the birds and laid them on a cloth.

Don Victor took out one of his 'in-stru-ments', a

stethoscope, and listened to their little hearts. He called Fidelio over, and let him listen. All he could hear was a faint 'thump . . . thump . . . thump'.

Fidelio caressed their bedraggled plumage, and talked to them. 'Please don't die, quetzalitos; you know you promised me you would not. Please live!' He said this to each bird. He ran to get cool water, and together Don Victor and he gave them water to drink and sprinkled more on them.

Hours passed, and Fidelio still would not lie down to sleep. He could not leave his little quetzals. Some had recovered a bit. They jumped about, but very painfully. They were so dizzy. Five of them seemed as if they were still dead, although Don Victor said they were not—he could still hear their hearts thump with his stethoscope. Fidelio kept talking to them and sitting by them.

Gradually, slowly, the birds began to try to get up. Yes, they were recovering. Even the one who had seemed most dead tried to get up and stretch its wing, but it was too tired and, falling over again, it was gone. The others claimed his attention.

Fidelio quickly got some corn and made it into balls, and rolled them on the floor for the quetzalitos to play with just as he had done before. The tired birds stood, wobbly and uncertain, watching the corn roll by. With an effort, one ran after, and another, and another. Fidelio shouted for joy. 'The quetzalitos are not all going to die!' And, suddenly overcome with a great weariness, Fidelio went to sleep.

When he woke again it was night. Don Victor stood over him, shaking him gently. 'Come, Fidelio, we must go on again. The bad part is past. I have put the birds into their cages, and the Indians have already started down the road. Only one little quetzal is gone. You saved the rest. Come, Fidelio, the rest of the trip will be short. Tomorrow morning we will be where the white men live, and a motor-car will come for us at the foot of the mountain.'

Fidelio was still too tired to ask what a motor-car was. He just plodded on. Some time before dawn they began to climb a mountain which Don Victor called Montaña de Mico Quemada—the mountain of the burnt monkey. Around and around they went, climbing until the long valley through which they had walked was left behind.

By dawn they were on top of the mountain, and the birds were fully recovered in the cool, high mountain air. When Fidelio went to feed them, they ate and ate, just as they had in Portillo Grande.

'Look, Don Victor, the birds are well again.'

'Yes,' he replied, 'I believe this is what they wanted. Well, we did it. We are not through yet, but the worst is over.' Even Doña Cristina looked less drawn about the eyes, and smiled as she looked at the lively birds and the proud Fidelio.

They went on, the Indians and Fidelio sleeping as they walked, and Doña Cristina asleep in her saddle.

Finally, Fidelio opened his tired eyes. Below him lay the Ulua valley, where white men raised bananas. Far below were many houses. Down, down they went to meet a road at the bottom. The Indians put down their cages in the shade of a tree and went to sleep, and Fidelio followed their example. He dreamed about the Plumed Serpent and about the quetzals. He thought he heard the Sisimiki roaring at him, making so much noise that he wanted to run away. He opened his eyes, but the roaring had not stopped. Something with big wheels standing beside him was making that awful noise. Although he wanted to get up and run, he was too exhausted. He heard strange voices, but he could not open his eyes wide enough to look. His brain was asleep. He felt someone lift him and put him beside the other Indians on the floor of whatever this was—then it came through his muddled mind that Don Victor had said a motor-car would meet them. Perhaps this was a motor-car.

As he sank off to sleep he could hear Don Victor saying to Doña Cristina, 'We made it, Cristina, we made it. The quetzals are happy and well, and we have broken a myth of four hundred years. The quetzal has lived in captivity.'

Fidelio sank into the deepest sleep he had ever had.

# THE QUETZAL MYTH IS BROKEN

Fidelio's ears still roared. Was he dreaming? He tried to open his eyes, but they were too heavy. Was he still in the motor-car that roared like the Sisimiki? He finally dragged one eye open. He was in a house, sleeping on the floor. A pillow was under his head. The roar did not cease, until at last Fidelio was wide awake. He got up.

Why, there were the quetzalitos! All sitting on their perches looking cool and nice in spite of a few feathers missing. Cold air was blowing on them. He looked around to see where a cold wind could come from in such a hot place. On the table was something whirling around, making this wind. He held his hand close to it.

'Look out, Fidelio, look out! Don't go near that fan!'

Fidelio jumped back at the command as if he had run into a thorn tree.

'*Buenos dias*, Don Victor,' he said, recovering from his scare. 'What did I do wrong?'

'The electric fan which you see whirling around making the breeze for the quetzalitos has blades that could cut your hand off if you got it too close. You must not go near it.'

'What is a *lec trick-fane* for?'

'Don't you see? I have put some chunks of ice in

front of it and the cold wind blows across the quetzalitos. It is as if they were still up on the mountain.
You see, they do not feel the heat this way.'

'What is *ice*, Don Victor?' Fidelio wanted to know.

'There in that bowl. When it grows very cold, the
water becomes hard and thick—it freezes.'

'*Freezes?*' asked Fidelio in wonderment.

Don Victory laughed, for it was very difficult to
explain more new words to Fidelio. 'Come here,' he
said, 'touch this, this is ice.' He put Fidelio's hands on
the ice.

Fidelio drew back. 'It is hot!'

'No,' Don Victor said, 'it is cold. Now you take care
to see that the quetzalitos eat well, for tomorrow
morning we are going to the coast where the ships are.'

'Do we have to walk some more?' asked Fidelio
sadly.

'No, no,' laughed Don Victor, 'our walking days are
done, for the time being. No more walking. We will
ride down tomorrow.'

When Don Victor had gone, Fidelio busied himself
with the birds, feeding them—giving them their 'cood
leefer ool' and at last, taking them on his fingers, he
tried to arrange their plumage  It was all out of shape
from the bouncing about in the cages during the long
walk. They had lived through the long expedition, and
Fidelio, tears coming to his eyes in his happiness,
talked to the birds.

'You see, quetzalitos, you were all good birds. You
did just what Fidelio asked. You lived to come here.
Now you must live some more, when Don Victor
takes you back on the ship. Here, take another piece

of avocado.' The birds gulped at the delicious green fruit, and jumped back to Fidelio for more.

Later Don Victor came in, looking worriedly at a piece of paper. 'Something awful has happened, Fidelio. I have just received a telegram. I cannot go back with the birds. Too much time will be lost. Who can feed the birds properly? No one will know how. I can't let this happen now. The birds will be on board five days. Someone must take care of them.' He paced back and forth, angrily.

Fidelio shook his head regretfully for a moment, and then his eyes gleamed. 'Don Victor,' he said timidly, 'there's me!'

Don Victor seemed not to have heard him. He repeated, a little louder, 'There's me, Don Victor. I could feed the birds.'

Turning quickly, and looking hard at the eager little boy, Don Victor exclaimed, 'Of course! Of course there's you! I had forgotten. Would you like to go on the ship and see the white man's city—big buildings—trains—motor-cars? You can stay there five days, and come back with the ship.'

Fidelio glowed at the thought. The ship—to ride upon it!

'I will speak to the captain,' continued Don Victor, 'and I think it can be arranged. Fine! Fine!' He struck his fist into the palm of his hand. 'Why, of course, who could take care of the birds better than you? I will send one of the Indians back to Portillo Grande to tell Old Chico about it. The other can stay here until you return and then go back with you to your home.'

The plan was soon completed. The next day they

went down to the coast. Fidelio has seen so many wonders that nothing startled him until he noticed the driver of the car, a very black boy, extremely bow-legged. He stood looking at the boy.

'What are you staring at, Fidelio?' Don Victor shouted. 'Come on, climb into the car. Oh, you are looking at the man with the bow-legs. What is so strange about that?'

'I was just thinking,' said Fidelio, reflectively, 'if his mamma had given him some "cood leefer ool" when he was little, he would not have crooked legs.'

Doña Cristina and Don Victor laughed and laughed, and the car started off down the road.

At noontime, after a nap, Fidelio woke and found that they were nearing the coast. A strange, refreshing, vibrant feeling was in the atmosphere. He had never known it, and his nostrils quivered as he drank in this new, fresh smell. He saw bananas, thousands of bananas, lying in piles—then, as the motor-car rounded a curve, he saw a ship. A big white ship. He had never thought it was like this—as big as this! Old Chico had said big—but this—he had not supposed anything could be so large. It was like a huge house that floated. He was sure it was a ship at once, because he saw it moving up and down with the waves.

'Well, there is the ship,' said Don Victor. 'That is the one you will go on. We get off here. Now carry the quetzals into this office and you stay beside them. Do not leave them for one little minute.'

Soon the word went around the whole wharf. All the men that were working heard that these people had brought quetzal birds alive. They, too, had heard of the quetzals, but had never seen a real one. They crowded close to Fidelio.

'Why, look,' said one, 'he is a little Indio.' Someone playfully pulled Fidelio's long pigtail that hung over his shoulder. Fidelio just gave them a look of

disdain, and passed into the house, carrying his precious burden.

Later, he met the manager of the docks, a pleasant, smiling white man, who talked nicely to him and laughed often as Don Victor did. 'So, you are Fidelio! And you're going back with the birds. How old are you? Twelve? Why, you are only a little runt.'

'What is a runt, señor?' asked Fidelio, respectfully. 'Isn't that what a pig does when you kick him?'

'No, no, that is "grunt"—but I was only joking. I thought you were a bigger boy. Don Victor tells me that you would like to go on this ship to take care of the birds. This is not done often, but I think we can sign you on as a mess-boy. Your duties will be to take care of the birds. Now you go on and give the birds their "cood leefer ool"!' The big man burst out into laughter again.

Don Victor was having another cage made, a much bigger one. 'It seems to me,' he confided to Fidelio, 'that for the past four months all I have done is make cages.' A carpenter was called in this time by the manager of the port, and he was making a cage of thin wood to be covered with mosquito netting. It had to be big, for the quetzalitos could not be taken out to exercise on board ship, as they might fly overboard.

The carpenter worked all day and far into the night. He was very slow, and Don Victor, angered at the time the man had taken, spoke to him about it. Fidelio, too, wondered as he heard the question. The carpenter said slowly, 'Señor, I have not worked fast because I have been thinking. I cannot work and think at the same time. I have said to myself—these birds are not quetzals. Why do I say this? Well, for all time they say that when you touch a quetzal it dies. Now, this little Indio picks up the birds and plays with them. They do not die. Therefore they are not quetzals. It was of this I have been thinking.'

'Well,' said Don Victor, drily, 'let me save your time and think for you. They are quetzals, and I want them in this cage tonight.'

The next day the boat sailed. Fidelio was too excited to eat. Doña Cristina had gone to town and bought new clothes for him, as well as a new hat. They had also taken him to a barber's shop and had his long pigtail cut off, to prevent the boys from making fun of him, or pulling his hair. Don Victor put a whole

handful of silver dollars into his pockets, and went over
with him again and again just what to do for the birds.

'If the birds do not seem to be feeling well, and do
not want to eat, give them some mineral oil! No, no,
not "cood leefer"—you give them that every morning.
The mineral oil is the thick white oil in the big orange
bottle. Then you feed them again. Right?'

Fidelio understood, but only half heard, for he was
anxious to go aboard the ship. At last the cage had
been put on board, on the top deck near the wheel-
house, where they steered the boat. Fidelio hung on
the rail of the boat, watching every movement. Now
the lines were cast off. The whistle sounded! Fidelio
jumped. Now the ship was moving, slowly, slowly.

Down there on the dock were Don Victor and Doña
Cristina, waving to him. He took off his hat and waved
back. He watched them as the boat began to turn and
shape its course for the open sea. Don Victor and

Doña Cristina still stood on the shore, waving, and he waved and watched until they were tiny specks.

The sun was sinking, making the sky a glowing red. The last golden rays flashed on the green plumage of the quetzalitos, and the feathers gave out shimmering gold sparks just as Fidelio remembered seeing in the forest when the great long-tailed quetzal flew. He stood rapt, his big brown eyes very large and full, as he thought he heard Old Chico saying—something he had heard long ago——

'And then the Plumed Serpent built himself a raft from the skins of serpents and set sail on the ocean to the East.'

*How to pronounce some of the Spanish and Indian
words used in this story*

Aguacatillo—*ah-wah-cot-ee'yo*
Ampusay—*am-poo-sigh'*
Buenas tardes—*boo-ayn'as tar'days*
Cetzontli—*set-sont'lee*
Chico—*Cheek'o*
Cuyamapa—*Koo-yah-mah'pah*
Fidelio—*Fee-day'lee-o*
Hernandez—*Er-nahn'dez*
Hilguero—*Eel-gway'ro*
Ignacio—*Een-yaht'zee-o*
Itzapalapan—*Eet-sah-pah'lah-pan*
Jicaque—*Hee-kah'kee*
La Peña—*la payn'ya*
Miguel—*Mee-gell'*
Milpas—*meel'pass*
Moctezuma—*Mah-tuh-zoo'ma* (sometimes spelled Monte-
   zuma)
Montaña—*mon-tahn'ya*
Padre—*pah'dray*
Pedro—*Pay'dro*
Pepe—*Peppy*
Portillo Grande—*Por-tee'yo Grahn'day*
Pujante—*poo-hahn'tay*
Quetzal—*ketzaal'*
Quetzalcoatl—*ket-zal-co-what'l*
Quetzalito—*ket-zal-eet'o*
Quién sabe—*kee-yen' sah'bay*
Santiago—*sahn-tee-yah'go*
Si—*see*
Sisimiki—*See-see-mee'kee*
Tejon—*tee-hone'*
Tortillas—*tor-tee'ya*
Tumecas—*too-may'cass*